Electromagnetic Induction and Maxwell Equation

Electromagneic Induction and Maxwell Equation

Maharshi A. Mehta

AH

Authors Nation

New Delhi-110078 (India)

ELECTROMAGNEIC INDUCTION AND MAXWELL EQUATION

First Published 2018
ISBN : 978-81-934260-2-9

Price : 1500/-

Page: 216

Published in India by

Authors Nation
412 Metro View Apartment,
Sector 13 B, Dwarka
New Delhi 110078
E-mail : authorsnation@gmail.com

Printed at : Asian Offset Press Delhi

Preface

Electromagnetic induction (or sometimes just induction) is a process where a conductor placed in a changing magnetic field (or a conductor moving through a stationary magnetic field) causes the production of a voltage across the conductor. This process of electromagnetic induction, in turn, causes an electrical current - it is said to induce the current.

Michael Faraday is given credit for the discovery of electromagnetic induction in 1831, though some others had noted similar behavior in the years prior to this.

The formal name for the physics equation that defines the behavior of an induced electromagnetic field from the magnetic flux (change in a magnetic field) is Faraday's law of electromagnetic induction.

The process of electromagnetic induction works in reverse as well, so that a moving electrical charge generates a magnetic field. In fact, a traditional magnet is the result of the individual motion of the electrons within the individual atoms of the magnet, aligned so that the generated magnetic field is in a uniform direction. (In non-magnetic materials, the electrons move in such a way that the individual magnetic fields point in different directions, so they cancel each other out and the net magnetic field generated is negligible.)

Maxwell's equations are a set of partial differential equations that, together with the Lorentz force law, form the foundation of classical electromagnetism, classical optics, and electric circuits. They underpin all electric, optical and radio technologies, including power generation, electric motors, wireless communication, cameras, televisions, computers etc. Maxwell's equations describe how electric and magnetic fields are generated by charges, currents, and changes of each other. One important consequence of the equations is that they demonstrate how fluctuating electric and magnetic fields propagate at the speed of light. Known as electromagnetic radiation, these waves may occur at various wavelengths to produce a spectrum from radio waves to ?-rays. The equations are named after the physicist and mathematician James Clerk Maxwell, who between 1861 and 1862 published an early form of the equations, and first proposed that light is an electromagnetic phenomenon.

—Author

Contents

1

Introduction of Electromagnetism

HISTORY OF ELECTROMAGNETIC INDUCTION

The term electromagnetic induction refers to the generation of an electric current by passing a metal wire through a magnetic field. The discovery of electromagnetic induction in 1831 was preceded a decade earlier by a related discovery by Danish physicist Hans Christian Oersted (1777–1851). Oersted showed that an electric current produces a magnetic field. That is, if you place a simple magnetic compass near any of the electrical wires in your home that are carrying a current, you can detect a magnetic field around the wires. If an electric current can produce a magnetic field, physicists reasoned, perhaps the reverse effect could be observed as well. So they set out to generate an electric current from a magnetic field.

That effect was first observed in 1831 by English physicist Michael Faraday (1791–1867) and shortly thereafter by American physicist Joseph Henry (1797–1878). The principle on which the Faraday-Henry discovery is based is shown in the figure on page 762. A long piece of metal wire is wound around a metal bar. The two ends of the wire are connected to a galvanometer, an instrument used to measure electric current. The bar is then placed between the poles of a magnet.

ELECTROMAGNETISM

Electromagnetism is a branch of physics involving the study of the electromagnetic force, a type of physical interaction that occurs between electrically charged particles.

Electromagnetic induction (or sometimes just induction) is a process where a conductor placed in a changing magnetic field (or a conductor moving through a stationary magnetic field) causes the production of a voltage across the conductor. This process of electromagnetic induction, in turn, causes an electrical current - it is said to induce the current.

DISCOVERY OF ELECTROMAGNETIC INDUCTION

Electromagnetic induction was discovered independently by Michael Faraday in 1831 and Joseph Henry in 1832. Faraday was the first to publish the results of his experiments. In Faraday's first experimental demonstration of electromagnetic induction (August 29, 1831), he wrapped two wires around opposite sides of an iron ring (torus) (an arrangement similar to a modern toroidal transformer). Based on his assessment of recently discovered properties of electromagnets, he expected that when current started to flow in one wire, a sort of wave would travel through the ring and cause some electrical effect on the opposite side. He plugged one wire into a galvanometer, and watched it as he connected the other wire to a battery. Indeed, he saw a transient current (which he called a "wave of electricity") when he connected the wire to the battery, and another when he disconnected it. This induction was due to the change in magnetic flux that occurred when the battery was connected and disconnected. Within two months, Faraday had found several other manifestations of electromagnetic induction. For example, he saw transient currents when he quickly slid a bar magnet in and out of a coil of wires, and he generated a steady (DC) current by rotating a copper disk near the bar magnet with a sliding electrical lead ("Faraday's disk").

Michael Faraday explained electromagnetic induction using a concept he called lines of force. However, scientists at the time widely rejected his theoretical ideas, mainly because they were not formulated mathematically. An exception was James Clerk Maxwell, who in 1861-2 used Faraday's ideas as the basis of his quantitative electromagnetic theory. In Maxwell's papers, the time-varying aspect of electromagnetic induction is expressed as a differential equation which Oliver Heaviside referred to as Faraday's law even though it is different from the original version of Faraday's law, and does not describe motional EMF. Heaviside's version (see Maxwell–Faraday equation below) is the form recognized today in the group of equations known as Maxwell's equations.

Michael Faraday is given credit for the discovery of electromagnetic induction in 1831, though some others had noted similar behavior in the years prior to this.

The formal name for the physics equation that defines the behavior of an induced electromagnetic field from the magnetic flux (change in a magnetic field) is Faraday's law of electromagnetic induction.

The process of electromagnetic induction works in reverse as well, so that a moving electrical charge generates a magnetic field. In fact, a traditional magnet is the result of the individual motion of the electrons within the individual atoms of the magnet, aligned so that the generated magnetic field is in a uniform direction. (In non-magnetic materials, the electrons move in such a way that the individual magnetic fields point in different directions, so they cancel each other out and the net magnetic field generated is negligible.)

Originally, electricity and magnetism were considered to be two separate forces. This view changed, however, with the publication of James Clerk Maxwell's 1873 A Treatise on Electricity and Magnetism in which the interactions of positive and negative charges were shown to be mediated by one force. There are four main effects resulting from these interactions, all of which have been clearly demonstrated by experiments:

1. Electric charges attract or repel one another with a force inversely proportional to the square of the distance between them: unlike charges attract, like ones repel.
2. Magnetic poles (or states of polarization at individual points) attract or repel one another in a manner similar to positive and negative charges and always exist as pairs: every north pole is yoked to a south pole.
3. An electric current inside a wire creates a corresponding circumferential magnetic field outside the wire. Its direction (clockwise or counter-clockwise) depends on the direction of the current in the wire.
4. A current is induced in a loop of wire when it is moved toward or away from a magnetic field, or a magnet is moved towards or away from it; the direction of current depends on that of the movement.

André-Marie Ampère

While preparing for an evening lecture on 21 April 1820, Hans Christian Ørsted made a surprising observation. As he was setting up his materials, he noticed a compass needle deflected away from magnetic north when the electric current from the battery he was using was switched on and off. This deflection convinced him that magnetic fields radiate from all sides of a wire carrying an electric current, just as light and heat do, and that it confirmed a direct relationship between electricity and magnetism.

Michael Faraday

At the time of discovery, Ørsted did not suggest any satisfactory explanation of the phenomenon, nor did he try to represent the phenomenon in a mathematical framework. However, three months later he began more intensive investigations. Soon thereafter he published his findings, proving that an electric current produces a magnetic field as it flows through a wire. The CGS unit of magnetic induction (oersted) is named in honor of his contributions to the field of electromagnetism.

James Clerk Maxwell

His findings resulted in intensive research throughout the scientific community in electrodynamics. They influenced French physicist André-Marie Ampère's developments of a single mathematical form to represent the magnetic forces between current-carrying conductors. Ørsted's discovery also represented a major step toward a unified concept of energy.

This unification, which was observed by Michael Faraday, extended by James Clerk Maxwell, and partially reformulated by Oliver Heaviside and Heinrich Hertz, is one of the key accomplishments of 19th century mathematical physics. It has had far-reaching consequences, one of which was the understanding of the nature of light. Unlike what was proposed by electromagnetic theory of that time, light and other electromagnetic waves are at present seen as taking the form of quantized, self-propagating oscillatory electromagnetic field disturbances called photons. Different frequencies of oscillation give rise to the different forms of electromagnetic radiation, from radio waves at the lowest frequencies, to visible light at intermediate frequencies, to gamma rays at the highest frequencies.

Ørsted was not the only person to examine the relationship between electricity and magnetism. In 1802, Gian Domenico Romagnosi, an Italian legal scholar, deflected a magnetic needle using a Voltaic pile. The factual setup of the experiment is not completely clear, so if current flew across the needle or not. An account of the discovery was published in 1802 in an Italian newspaper, but it was largely overlooked by the contemporary scientific community, because Romagnosi seemingly did not belong to this community.

FUNDAMENTAL FORCES

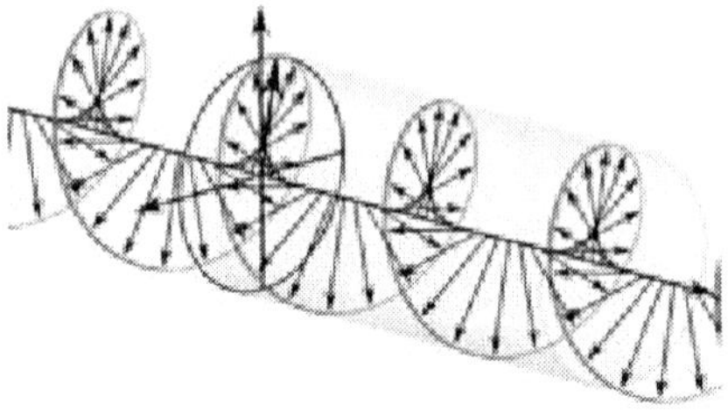

Representation of the electric field vector of a wave of circularly polarized electromagnetic radiation.

The electromagnetic force is one of the four known fundamental forces. The other fundamental forces are:

- the weak nuclear force, which binds to all known particles in the Standard Model, and causes certain forms of radioactive decay. (In particle physics though, the electroweak interaction is the unified description of two of the four known fundamental interactions of nature: electromagnetism and the weak interaction);
- the strong nuclear force, which binds quarks to form nucleons, and binds nucleons to form nuclei
- the gravitational force.

All other forces (e.g., friction, contact forces) are derived from these four fundamental forces (including momentum which is carried by the movement of particles).

The electromagnetic force is responsible for practically all phenomena one encounters in daily life above the nuclear scale, with the exception of gravity. Roughly speaking, all the forces involved in interactions between atoms can be explained by the electromagnetic force acting between the electrically charged atomic nuclei and electrons of the atoms. Electromagnetic forces also explain how these particles carry momentum by their movement. This includes the forces we experience in "pushing" or "pulling" ordinary material objects, which result from the intermolecular forces that act between the individual molecules in our bodies and those in the objects. The electromagnetic force is also involved in all forms of chemical phenomena.

A necessary part of understanding the intra-atomic and intermolecular forces is the effective force generated by the momentum of the electrons' movement, such that as electrons move between interacting atoms they carry momentum with them. As a collection of electrons becomes more confined, their minimum momentum necessarily increases due to the Pauli exclusion principle. The behaviour of matter at the molecular scale including its density is determined by the balance between the electromagnetic force and the force generated by the exchange of momentum carried by the electrons themselves.

CLASSICAL ELECTRODYNAMICS

In 1600, William Gilbert proposed, in his De Magnete, that electricity and magnetism, while both capable of causing attraction and repulsion of objects, were distinct effects. Mariners had noticed that lightning strikes had the ability to disturb a compass needle. The link between lightning and electricity was not confirmed until Benjamin Franklin's proposed experiments in 1752. One of the first to discover and publish a link between man-made electric current and magnetism was Romagnosi, who in 1802 noticed that connecting a wire across a voltaic pile deflected a nearby compass needle. However, the effect did not become widely known until 1820, when Ørsted performed a similar experiment. Ørsted's work influenced Ampère to produce a theory of electromagnetism that set the subject on a mathematical foundation.

A theory of electromagnetism, known as classical electromagnetism, was developed by various physicists during the period between 1820 and 1873 when it culminated in the publication of a treatise by James Clerk Maxwell, which unified the preceding developments into a single theory and discovered the electromagnetic nature of light. In classical electromagnetism, the behavior of the electromagnetic field is described by a set of equations known as Maxwell's equations, and the electromagnetic force is given by the Lorentz force law.

One of the peculiarities of classical electromagnetism is that it is difficult to reconcile with classical mechanics, but it is compatible with special relativity. According to Maxwell's equations, the speed of light in a vacuum is a universal constant that is dependent only on the electrical permittivity and magnetic permeability of free space. This violates Galilean invariance, a long-standing cornerstone of classical mechanics. One way to reconcile the two theories (electromagnetism and classical mechanics) is to assume the existence of a luminiferous aether through which the light propagates. However, subsequent experimental efforts failed to detect the presence of the aether. After important contributions of Hendrik Lorentz and Henri Poincaré, in 1905, Albert Einstein solved the problem with the introduction of special relativity, which replaced classical kinematics with a new theory of kinematics compatible with classical electromagnetism. (For more information, see History of special relativity.)

In addition, relativity theory implies that in moving frames of reference a magnetic field transforms to a field with a nonzero electric component and conversely, a moving electric field transforms to a nonzero magnetic component, thus firmly showing that the phenomena are two sides of the same coin. Hence the term "electromagnetism".

CHAOTIC AND EMERGENT PHENOMENA

The mathematical models used in classical electromagnetism, quantum electrodynamics (QED) and the standard model all view the electromagnetic force as a linear set of equations. In these theories electromagnetism is a U(1) gauge theory, whose topological properties do not allow any complex nonlinear interaction of a field with and on itself. For example, in the QED vacuum the field fluctuates randomly as a consequence of the uncertainty principle but these fluctuations cancel each other out with no overall observable effect. However, there are many observed nonlinear physical electromagnetic phenomena such as Aharonov–Bohm(AB) and Altshuler–Aronov–Spivak (AAS) effects, Berry, Aharonov– Anandan, Pancharatnam and Chiao–Wuphase rotation effects, Josephson effect, Quantum Hall effect, the de Haas–van Alphen effect, the Sagnac effect and many other physically observable phenomena which would indicate that the electromagnetic potential field has real physical meaning rather than being a mathematical artifact and therefore an all encompassing theory would not confine electromagnetism as a local force as is currently done, but as a SU(2) gauge theory or higher geometry. Higher symmetries allow for nonlinear, aperiodic behaviour which manifest as a variety of complex non-equilibrium phenomena that do not arise in the linearised U(1) theory, such as multiple stable states, symmetry breaking, chaos and emergence. In higher symmetry groups, the electromagnetic field is not a calm, randomly fluctuating, passive substance, but can at times be viewed as a turbulent virtual plasma that can have complex vortices, entangled states and a rich nonlinear structure.

What are called Maxwell's equation's today are in fact a simplified version of the original equations reformulated by Heaviside, FitzGerald, Lodge and Hertz. The original equations used Hamilton's more expressive quaternion notation, a kind of Clifford algebra, which fully subsumes the standard Maxwell vectorial equations largely used today. In the late 1880s there was a debate over the relative merits of vector analysis and quaternions. According to Heaviside the electromagnetic potential field was purely metaphysical, an arbitrary mathematical fiction, that needed to be "murdered". It was concluded that there was no need for the greater physical insights provided by the quaternions if the theory was purely local in nature. Local vector analysis has become the dominant way of using Maxwell's equations ever since. However, this strictly vectorial approach has led to a restrictive topological understanding in some areas of electromagnetism, for example, a full understanding of the energy transfer dynamics in Tesla's oscillator-shuttle-circuit can be achieved only in quaternionic algebra or higher SU(2) symmetries. It has often been argued that quaternions are not compatible with special relativity, but multiple papers have shown ways of incorporating relativity

Plasma vortices give rise to solar flares on the sun

A good example of nonlinear electromagnetics is in high energy dense plasmas, where vortical phenomena occur which seemingly violate the second law of thermodynamics by increasing the energy gradient within the electromagnetic field and violate Maxwell's laws by creating ion currents which capture and concentrate their own and surrounding magnetic fields. In particular Lorentz force law, which elaborates Maxwell's equations is violated by these force free vortices. These apparent violations are due to the fact that the traditional conservation laws in classical and quantum electrodynamics (QED) only display linear U(1) symmetry (in particular, by the extended Noether theorem, conservation laws such as the laws of thermodynamicsneed not always apply to dissipative systems, which are expressed in gauges of higher symmetry). The second law of thermodynamics states that in a closed linear system entropy flow can be only be positive (or exactly zero at the end of a cycle). However, negative entropy (i.e. increased order, structure or self-organisation) can spontaneously appear in an open nonlinear thermodynamic system that is far from equilibrium, so long as this emergent order accelerates the overall flow of entropy in the total system.

Below its critical temperature, a superconductor becomes perfectly diamagnetic and excludes sufficiently weak magnetic fields from passing through it. This is called the Meissner effect which is described by the London equations.

Given the complex and adaptive behaviour that arises from nonlinear systems considerable attention in recent years has gone into studying a new class of phase transitions which occur at absolute zero temperature. These are quantum phase transitions which are driven by electromagnetic field fluctuations as a consequence of zero-point energy A good example of a spontaneous phase transition that are attributed to zero-point fluctuations can be found in superconductors. Superconductivity is one of the best known empirically quantified macroscopic electromagnetic phenomena whose basis is recognised to be quantum mechanical in origin. The behaviour of the electric and magnetic fields under superconductivity is governed by the London equations. However, it has been questioned in a series of journal articles whether the quantum mechanically canonised London equations can be given a purely classical derivation.Bostick for instance, has claimed to show that the London equations do indeed have a classical origin that applies to superconductors and to some collisionless plasmas as well. In particular it has been asserted that the Beltrami vortices in the plasma focus display the same paired flux-tube morphology as Type II superconductors. Others have also pointed out this connection, Fröhlich has shown that the hydrodynamic equations of compressible fluids, together with the London equations, lead to a macroscopic parameter ({\displaystyle \mu }electric charge density / mass density), without involving either quantum phase factors or Planck's constant. In essence, it has been asserted that Beltrami plasma vortex structures are able to at least simulate the morphology of Type I and Type II superconductors. This occurs because the "organised" dissipative energy of the vortex configuration comprising the ions and electrons far exceeds the "disorganised" dissipative random thermal energy. The transition from disorganised fluctuations to organised helical structures is a phase transition involving a change in the condensate's energy (i.e. the ground state or zero-point energy) but without any associated rise in temperature. This is an example of zero-point energy having multiple stable states (see Quantum phase transition, Quantum critical point, Topological degeneracy, Topological order) and where the overall system structure is independent of a reductionist or deterministic view, that "classical" macroscopic order can also causally affect quantum phenomena.

QUANTITIES AND UNITS

Electromagnetic units are part of a system of electrical units based primarily upon the magnetic properties of electric currents, the fundamental SI unit being the ampere. The units are:

- ampere (electric current)
- coulomb (electric charge)
- farad (capacitance)
- henry (inductance)
- ohm (resistance)
- siemens (conductance)
- tesla (magnetic flux density)
- volt (electric potential)
- watt (power)
- weber (magnetic flux)

In the electromagnetic cgs system, electric current is a fundamental quantity defined via Ampère's law and takes the permeability as a dimensionless quantity (relative permeability) whose value in a vacuum is unity. As a consequence, the square of the speed of light appears explicitly in some of the equations interrelating quantities in this system.

SI electromagnetism units				
Symbol	**Name of Quantity**	**Derived Units**	**Unit**	**Base Units**
I	electric current	ampere (SI base unit)	A	A (= W/V = C/s)
Q	electric charge	coulomb	C	A·s
$U, \Delta V, \Delta\varphi; E$	potential difference; electromotive force	volt	V	kg·m·s·A (= J/C)
$R; Z; X$	electric resistance; impedance; reactance	ohm	Ω	kg·m·s·A (= V/A)
ρ	resistivity	ohm metre	Ω·m	kg·m·s·A
P	electric power	watt	W	kg·m·s (= V·A)
C	capacitance	farad	F	kg·m·s·A (= C/V)
E	electric field strength	volt per metre	V/m	kg·m·s·A (= N/C)
D	electric displacement field	coulomb per square metre	C/m	A·s·m
ε	permittivity	farad per metre	F/m	kg·m·s·A
χ_e	electric susceptibility	(dimensionless)	–	–
$G; Y; B$	conductance; admittance; susceptance	siemens	S	kg·m·s·A (= Ω)
κ, γ, σ	conductivity	siemens per metre	S/m	kg·m·s·A
B	magnetic flux density, magnetic induction	tesla	T	kg·s·A (= Wb/m = N·A·m)
	magnetic flux	weber	Wb	kg·m·s·A (= V·s)
H	magnetic field strength	ampere per metre	A/m	A·m
L, M	inductance	henry	H	kg·m·s·A (= Wb/A = V·s/A)
μ	permeability	henry per metre	H/m	kg·m·s·A
χ	magnetic susceptibility	(dimensionless)	–	–
J	current density	ampere per square metre	A/m	C·m·s

Formulas for physical laws of electromagnetism (such as Maxwell's equations) need to be adjusted depending on what system of units one uses. This is because there is no one-to-one correspondence between electromagnetic units in SI and those in CGS, as is the case for mechanical units. Furthermore, within CGS, there are several plausible choices of electromagnetic units, leading to different unit "sub-systems", including Gaussian, "ESU", "EMU", and Heaviside–Lorentz. Among these choices, Gaussian units are the most common today, and in fact the phrase "CGS units" is often used to refer specifically to CGS-Gaussian units.

COMPUTATIONAL ELECTROMAGNETICS

computational electrodynamics or electromagnetic modeling is the process of modeling the interaction of electromagnetic fields with physical objects and the environment.

It typically involves using computationally efficient approximations to Maxwell's equations and is used to calculate antenna performance, electromagnetic compatibility, radar cross section and electromagnetic wave propagation when not in free space.

A specific part of computational electromagnetics deals with electromagnetic radiation scattered and absorbed by small particles.

MECHANICAL PROPERTIES OF MATTER

The first constitutive equation (constitutive law) was developed by Robert Hooke and is known as Hooke's law. It deals with the case of linear elastic materials. Following this discovery, this type of equation, often called a "stress-strain relation" in this example, but also called a "constitutive assumption" or an "equation of state" was commonly used. Walter Noll advanced the use of constitutive equations, clarifying their classification and the role of invariance requirements, constraints, and definitions of terms like "material", "isotropic", "aeolotropic", etc. The class of "constitutive relations" of the form stress rate = f (velocity gradient, stress, density) was the subject of Walter Noll's dissertation in 1954 under Clifford Truesdell.

In modern condensed matter physics, the constitutive equation plays a major role. See Linear constitutive equations and Nonlinear correlation functions.

CONSTITUTIVE EQUATIONS IN ELECTROMAGNETISM AND RELATED AREAS

In both classical and quantum physics, the precise dynamics of a system form a set of coupled differential equations, which are almost always too complicated to be solved exactly, even at the level of statistical mechanics. In the context of electromagnetism, this remark applies to not only the dynamics of free charges and currents (which enter Maxwell's equations directly), but also the dynamics of bound charges and currents (which enter Maxwell's equations through the constitutive relations). As a result, various approximation schemes are typically used.

For example, in real materials, complex transport equations must be solved to determine the time and spatial response of charges, for example, the Boltzmann equation or the Fokker–Planck equation or the Navier-Stokes equations. For example, see magnetohydrodynamics, fluid dynamics, electrohydrodynamics, superconductivity, plasma modeling. An entire physical apparatus for dealing with these matters has developed. See for example, linear response theory, Green–Kubo relations and Green's function (many-body theory).

These complex theories provide detailed formulas for the constitutive relations describing the electrical response of various materials, such as permittivities, permeabilities, conductivities and so forth.

It is necessary to specify the relations between displacement field D and E, and the magnetic H-field H and B, before doing calculations in electromagnetism (i.e. applying Maxwell's macroscopic equations). These equations specify the response of bound charge and current to the applied fields and are called constitutive relations.

CALCULATION OF CONSTITUTIVE RELATIONS

The theoretical calculation of a material's constitutive equations is a common, important, and sometimes difficult task in theoretical condensed-matter physics and materials science. In general, the constitutive equations are theoretically determined by calculating how a molecule responds to the local fields through the Lorentz force. Other forces may need to be modeled as well such as lattice vibrations in crystals or bond forces. Including all of the forces leads to changes in the molecule which are used to calculate P and M as a function of the local fields.

The local fields differ from the applied fields due to the fields produced by the polarization and magnetization of nearby material; an effect which also needs to be modeled. Further, real materials are not continuous media; the local fields of real materials vary wildly on the atomic scale. The fields need to be averaged over a suitable volume to form a continuum approximation.

These continuum approximations often require some type of quantum mechanical analysis such as quantum field theory as applied to condensed matter physics. See, for example, density functional theory, Green–Kubo relations and Green's function.

A different set of homogenization methods (evolving from a tradition in treating materials such as conglomerates and laminates) are based upon approximation of an inhomogeneous material by a homogeneous effective medium (valid for excitations with wavelengths much larger than the scale of the inhomogeneity).

The theoretical modeling of the continuum-approximation properties of many real materials often rely upon experimental measurement as well. For example, å of an insulator at low frequencies can be measured by making it into a parallel-plate capacitor, and å at optical-light frequencies is often measured by ellipsometry.

Thermoelectric and electromagnetic properties of matter

These constitutive equations are often used in crystallography, a field of solid-state physics.

ELECTROMAGNETIC INDUCTION

Is the production of a potential difference (voltage) acrossa conductor when it is exposed to a varying magnetic field. Michael Faraday is generally credited with the discovery of induction in 1831 though it mayhave been anticipated by the work of Francesco Zantedeschi in 1829.

ELECTROMAGNETIC HYPERSENSITIVITY

Electromagnetic hypersensitivity (EHS) is a claimed sensitivity to electromagnetic fields, to which negative symptoms are attributed. EHS has no scientific basis and is not a recognised medical diagnosis. Claims are characterized by a "variety of non-specific symptoms, which afflicted individuals attribute to exposure to electromagnetic fields".

Those who are self-described with EHS report adverse reactions to electromagnetic fields at intensities well below the maximum levels permitted by international radiation safety standards. The majority of provocation trials to date have found that such claimants are unable to distinguish between exposure and non-exposure to electromagnetic fields. A systematic review in 2005 showed no convincing scientific evidence for symptoms being caused by electromagnetic fields. Since then, several double-blind experimentshave shown that people who report electromagnetic hypersensitivity are unable to detect the presence of electromagnetic fields and are as likely to report ill health following a sham exposure as they are following exposure to genuine electromagnetic fields, suggesting the cause in these cases to be the nocebo effect.

A 2005 review by the UK Health Protection Agency and a 2006 systematic review each evaluated the evidence for various medical, psychological, behavioral, and alternative treatments for EHS and each found that the evidence-base was limited and not generalizable, but that the best evidence favored cognitive behavioural therapy. As of 2005, WHO recommended that people presenting with claims of EHS be evaluated to determine if they have a medical condition that may be causing the symptoms the person is attributing to EHS, that they have a psychological evaluation, and that the person's environment be evaluated for issues like air or noise pollution that may be causing problems.

Some people who feel they are sensitive to electromagnetic fields may seek to reduce their exposure or use alternative medicine. Government agencies have enforced false advertising claims against companies selling devices to shield against EM radiation.

SIGNS AND SYMPTOMS

There are no specific symptoms associated with claims of EHS and reported symptoms range widely between individuals. They include headache, fatigue, stress, sleep disturbances, skin prickling, burning sensations and rashes, pain and ache in muscles and many other health problems. In severe cases such symptoms can be a real and sometimes disabling problem for the affected person, causing psychological distress. There is no scientific basis to link such symptoms to electromagnetic field exposure.

The prevalence of some reported symptoms is geographically or culturally dependent and does not imply "a causal relationship between symptoms and attributed exposure".Many such reported symptoms overlap with other syndromes known as symptom-based conditions, functional somatic syndromes, and IEI (idiopathic environmental intolerance).

Those reporting electromagnetic hypersensitivity will usually describe different levels of susceptibility to electric fields, magnetic fields, and various frequencies of electromagnetic waves. Devices implicated include fluorescent and low-energy lights, mobile, cordless/portable phones, and WiFi. A 2001 survey found that people self-diagnosing as EHS related their symptoms most frequently to mobile phone base stations (74%), followed by mobile phones (36%), cordless phones (29%), and power lines (27%). Surveys of electromagnetic hypersensitivity sufferers have not been able to find any consistent pattern to these symptoms.

CAUSES

Most blinded conscious provocation studies have failed to show a correlation between exposure and symptoms, leading to the suggestion that psychological mechanisms play a role in causing or exacerbating EHS symptoms. In 2010, Rubin et al. published a follow-up to their 2005 review, bringing the totals to 46 double-blind experiments and 1175 individuals with self-diagnosed hypersensitivity. Both reviews found no robust evidence to support the hypothesis that electromagnetic exposure causes EHS, as have other studies.They also concluded that the studies supported the role of the nocebo effect in triggering acute symptoms in those with EHS.

Some other types of studies suggest evidence for symptoms at non-thermal levels of electromagnetic exposure. A review in 2010 of ten studies on neurobehavioral and cancer outcomes near cell phone base stations found eight with increased prevalence, including sleep disturbance and headaches. Since 1962, the microwave auditory effect or tinnitus has been shown from radio frequency exposure at levels below significant heating. Studies during the 1960s in Europe and Russia claimed to show effects on humans, especially the nervous system, from low energy RF radiation; the studies were disputed at the time.

Other studies on sensitivity have looked at therapeutic procedures using non-thermal electromagnetic exposure, genetic factors, an alteration in mast cells, oxidative stress, protein expression and voltage-gated calcium channels. Mercury release from dental amalgam and heavy metal toxicity have also been implicated in exposure effects and symptoms. Another line of study has been the nature of hyper-sensitivity or intolerance and the range of environmental exposures which may be related to it. Some 80% of people with self-diagnosed electromagnetic intolerance also claim intolerance to low levels of chemical exposure.

DIAGNOSIS

Electromagnetic hypersensitivity is not an accepted diagnosis; medically there is no case definition or clinical practice guideline and there is no specific test to identify it, nor is there an agreed-upon definition with which to conduct clinical research.

Complaints of electromagnetic hypersensitivity may mask organic or psychiatric illness. Diagnosis of those underlying conditions involves investigating and identifying possible known medical causes of any symptoms observed. It may require both a thorough medical evaluation to identify and tre..t any specific conditions that may be responsible for the symptoms, and a psychological evaluation to identify alternative psychiatric/psychological conditions that may be responsible or contribute to the symptoms.

Symptoms may also be brought on by imagining that exposure is causing harm, an example of the nocebo effect. Studies have shown that reports of symptoms are more closely associated with belief that one is being exposed than with any actual exposure.

MANAGEMENT

A 2006 systematic review and a 2005 review by the UK Health Protection Agency each evaluated the evidence for various medical, psychological, behavioral, and alternative treatments for EHS and each found that the evidence-base was limited and not generalizable. The conclusion of the 2006 review stated: "The evidence base concerning treatment options for electromagnetic hypersensitivity is limited and more research is needed before any definitive clinical recommendations can be made. However, the best evidence currently available suggests that cognitive behavioural therapy is effective for patients who report being hypersensitive to weak electromagnetic fields."

As of 2005, WHO recommended that people presenting with claims of EHS be evaluated to determine if they have a medical condition that may be causing the symptoms the person is attributing to EHS, that they have a psychological evaluation, and that the person's environment be evaluated for issues like air or noise pollution that may be causing problems.

PREVALENCE

The prevalence of claimed electromagnetic hypersensitivity has been estimated as being between a few cases per million to 5% of the population depending on the location and definition of the condition.

In 2002, a questionnaire survey of 2,072 people in California found that the prevalence of self-reported electromagnetic hypersensitivity within the sample group was 3% (95% CI 2.8–3.68%), with electromagnetic hypersensitivity being defined as "being allergic or very sensitive to getting near electrical appliances, computers, or power lines" (response rate 58.3%).

A similar questionnaire survey from the same year in Stockholm County (Sweden), found a 1.5% prevalence of self-reported electromagnetic hypersensitivity within the sample group, with electromagnetic hypersensitivity being defined as "hypersensitivity or allergy to electric or magnetic fields" (response rate 73%).

A 2004 survey in Switzerland found a 5% prevalence of claimed electromagnetic hypersensitivity in the sample group of 2,048.

In 2007, a UK survey aimed at a randomly selected group of 20,000 people found a prevalence of 4% for symptoms self-attributed to electromagnetic exposure.

A group of scientists also attempted to estimate the number of people reporting "subjective symptoms" from electromagnetic fields for the European Commission. In the words of a HPA review, they concluded that "the differences in prevalence were at least partly due to the differences in available information and media attention around electromagnetic hypersensitivity that exist in different countries. Similar views have been expressed by other commentators."

SOCIETY AND CULTURE

In 2010, a cell tower operator in South Africa revealed at a public meeting that the tower that nearby residents were blaming for their current EHS symptoms had been turned off over six weeks prior to the meeting, thus making it a highly unlikely cause of EHS symptoms.

In February 2014, the UK Advertising Standards Authority found that claims of harm from electromagnetic radiation, made in a product advertisement, were unsubstantiated and misleading.

People have filed lawsuits to try to win damages due to harm claimed from electromagnetic radiation. In 2012, a New Mexico judge dismissed a lawsuit in which one person sued his neighbor, claiming to have been harmed by EM radiation from his neighbor's cordless telephones, dimmer switches, chargers, Wi-Fi and other devices. The plaintiff brought the testimony of his doctor, who also believed she had EHS, and a person who represented himself as a neurotoxicologist; the judge found none of their testimony credible. In 2015, parents of a boy at a school in Southborough, Massachusetts alleged that the school's wi-fi was making the boy sick.

In November 2015, a depressed teenage girl in England committed suicide. Her suicide was attributed to EHS by her parents and taken up by tabloids and EHS advocates.

Some people who feel they are sensitive to electromagnetic fields self-treat by trying to reduce their exposure to electromagnetic sources by avoiding sources of exposure, disconnecting or removing electrical devices, shielding or screening of self or residence, and alternative medicine. In Sweden, some municipalities provide disability grants to people who claim to have EHS in order to have abatement work done in their homes even though the public health authority does not recognize EHS as an actual medical condition; towns in Halland do not provide such funds and this decision was challenged and upheld in court.

The United States National Radio Quiet Zone is an area where wireless signals are restricted for scientific research purposes, and some people who believe they have EHS have relocated there seeking relief.

Charles McGill, a fictional character on the television show Better Call Saul is depicted as experiencing the symptoms of EHS. In the episode Alpine Shepherd Boy, a skeptical doctor proves that Chuck's electromagnetic hypersensitivity may not be genuine by surreptitiously turning on the electronics in his hospital bed.

DEFINITION OF INDUCTION

1a : the act or process of inducting (as into office)

b : an initial experience : INITIATION

c : the formality by which a civilian is inducted into military service

2a (1) : inference of a generalized conclusion from particular instances compare DEDUCTION 2a

(2) : a conclusion arrived at by induction

b : mathematical demonstration of the validity of a law concerning all the positive integers by proving that it holds for the integer 1 and that if it holds for an arbitrarily chosen positive integer k, it must hold for the integer k + 1 — called also mathematical induction

3: a preface, prologue, or introductory scene especially of an early English play

4a : the act of bringing forward or adducing something (such as facts or particulars)

b : the act of causing or bringing on or about

c : the process by which an electrical conductor becomes electrified when near a charged body, by which a magnetizable body becomes magnetized when in a magnetic field or in the magnetic flux set up by a magnetomotive force, or by which an electromotive force is produced in a circuit by varying the magnetic field linked with the circuit

d : the inspiration of the fuel-air charge from the carburetor into the combustion chamber of an internal combustion engine

e : the process by which the fate of embryonic cells is determined (as by the action of adjacent cells) and morphogenetic differentiation brought about

FARADAY'S LAW OF INDUCTION

is a basic law of electromagnetism predicting how a magneticfield will interact with an electric circuit to produce anelectromotive force (EMF).It is thefundamental operating principle of transformers, inductors,and many typesof electrical motors, generators andsolenoids.

THEMAXWELL

Faraday equation is a generalisation of Faraday's law, and forms one of Maxwell's equations.

MAXWELL'S EQUATIONS

Are a set of partial differential equations that, together with the Lorentzforce law, form the foundation of classical electrodynamics, classical optics,and electriccircuits.These fields in turn underlie modern electrical and communications technologies.Maxwell's equations describe howelectric and magnetic fields are generated and altered byeach other and by charges and currents.They are named after the Scottish physicist andmathematician James Clerk Maxwell who published an early form of those equationsbetween 1861 and 1862.The equations have two major variants. The "microscopic" set of Maxwell's equations usestotal charge and total current, including the complicated charges and currents in materials atthe atomic scale; it has universal applicability, but may be unfeasible to calculate. The"macroscopic" set of Maxwell's equations defines two new auxiliary fields that describelarge-scale behavior without having to consider these atomic scale details, but it requires theuse of parameters characterizing the electromagnetic properties of the relevant materials.The term "Maxwell's equations" is often used for other forms of Maxwell's equations. For example, space-time formulations are commonly used in high energy and gravitationalphysics. These formulations defined on space-time,rather than space and time separatelyare manifestly

Compatible withspecial and general relativity.In quantum mechanics, versions of Maxwell's equations based on the electric and magnetic potentials are preferred.Since the mid-20th century, it has been understood that Maxwell's equations are not exactlaws of the universe, but are a classical approximation to the more accurate andfundamental theory of quantum electrodynamics. In most cases, though, quantum deviationsfrom Maxwell's equations are immeasurably small. Exceptions occur whenthe particle nature of light is important or for very strong electric fields.

If the wire is then wound into a coil, the magnetic field is greatly intensified producing a static magnetic field around itself forming the shape of a bar magnet giving a distinct North and South pole.

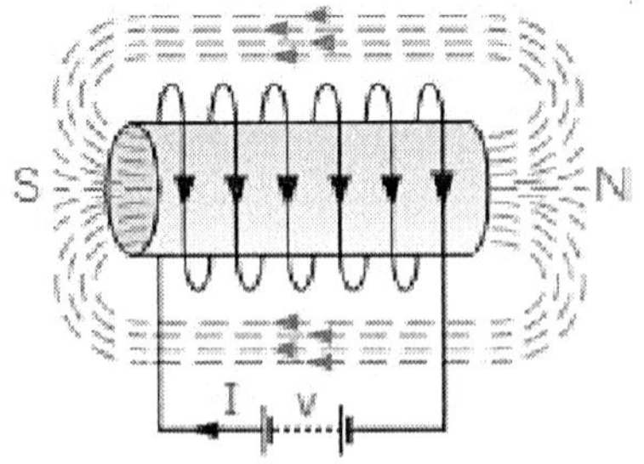

AIR-CORE HOLLOW COIL

The magnetic flux developed around the coil being proportional to the amount of current flowing in the coils windings as shown. If additional layers of wire are wound upon the same coil with the same current flowing through them, the static magnetic field strength would be increased.

Therefore, the magnetic field strength of a coil is determined by the ampere turns of the coil. With more turns of wire within the coil, the greater the strength of the static magnetic field around it.

But what if we reversed this idea by disconnecting the electrical current from the coil and instead of a hollow core we placed a bar magnet inside the core of the coil of wire. By moving this bar magnet "in" and "out" of the coil a current would be induced into the coil by the physical movement of the magnetic flux inside it.

Likewise, if we kept the bar magnet stationary and moved the coil back and forth within the magnetic field an electric current would be induced in the coil. Then by either moving the wire or changing the magnetic field we can induce a voltage and current within the coil and this process is known as Electromagnetic Induction and is the basic principal of operation of transformers, motors and generators.

Electromagnetic Induction was first discovered way back in the 1830's by Michael Faraday. Faraday noticed that when he moved a permanent magnet in and out of a coil or a single loop of wire it induced an ElectroMotive Force or emf, in other words a Voltage, and therefore a current was produced.

So what Michael Faraday discovered was a way of producing an electrical current in a circuit by using only the force of a magnetic field and not batteries. This then lead to a very important law linking electricity with magnetism, Faraday's Law of Electromagnetic Induction. So how does this work?.

When the magnet shown below is moved "towards" the coil, the pointer or needle of the Galvanometer, which is basically a very sensitive centre zero'ed moving-coil ammeter, will deflect away from its centre position in one direction only. When the magnet stops moving and is held stationary with regards to the coil the needle of the galvanometer returns back to zero as there is no physical movement of the magnetic field.

Likewise, when the magnet is moved "away" from the coil in the other direction, the needle of the galvanometer deflects in the opposite direction with regards to the first indicating a change in polarity. Then by moving the magnet back and forth towards the coil the needle of the galvanometer will deflect left or right, positive or negative, relative to the directional motion of the magnet.

ELECTROMAGNETIC INDUCTION BY A MOVING MAGNET

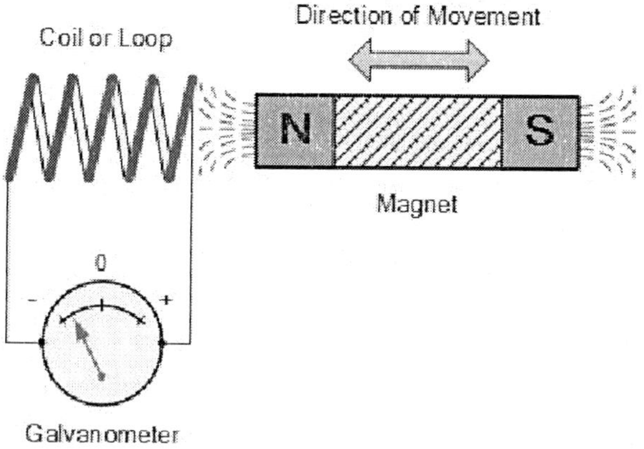

Likewise, if the magnet is now held stationary and ONLY the coil is moved towards or away from the magnet the needle of the galvanometer will also deflect in either direction. Then the action of moving a coil or loop of wire through a magnetic field induces a voltage in the coil with the magnitude of this induced voltage being proportional to the speed or velocity of the movement.

Then we can see that the faster the movement of the magnetic field the greater will be the induced emf or voltage in the coil, so for Faraday's law to hold true there must be "relative motion" or movement between the coil and the magnetic field and either the magnetic field, the coil or both can move.

FARADAY'S LAW OF INDUCTION

From the above description we can say that a relationship exists between an electrical voltage and a changing magnetic field to which Michael Faraday's famous law of electromagnetic induction states: "that a voltage is induced in a circuit whenever relative motion exists between a conductor and a magnetic field and that the magnitude of this voltage is proportional to the rate of change of the flux".

In other words, Electromagnetic Induction is the process of using magnetic fields to produce voltage, and in a closed circuit, a current.

So how much voltage (emf) can be induced into the coil using just magnetism. Well this is determined by the following 3 different factors.

- 1). Increasing the number of turns of wire in the coil – By increasing the amount of individual conductors cutting through the magnetic field, the amount of induced emf produced will be the sum of all the individual loops of the coil, so if there are 20 turns in the coil there will be 20 times more induced emf than in one piece of wire.
- 2). Increasing the speed of the relative motion between the coil and the magnet – If the same coil of wire passed through the same magnetic field but its speed or velocity is increased, the wire will cut the lines of flux at a faster rate so more induced emf would be produced.
- 3). Increasing the strength of the magnetic field – If the same coil of wire is moved at the same speed through a stronger magnetic field, there will be more emf produced because there are more lines of force to cut.

If we were able to move the magnet in the diagram above in and out of the coil at a constant speed and distance without stopping we would generate a continuously induced voltage that would alternate between one positive polarity and a negative polarity producing an alternating or AC output voltage and this is the basic principal of how a Generator works similar to those used in dynamos and car alternators.

In small generators such as a bicycle dynamo, a small permanent magnet is rotated by the action of the bicycle wheel inside a fixed coil. Alternatively, an electromagnet powered by a fixed DC voltage can be made to rotate inside a fixed coil, such as in large power generators producing in both cases an alternating current.

SIMPLE GENERATOR USING MAGNETIC INDUCTION

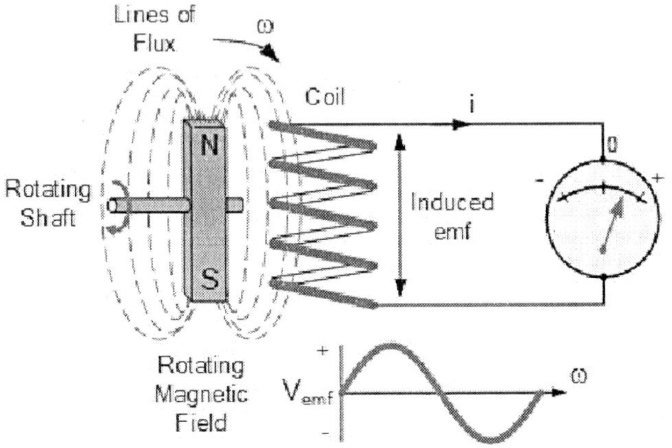

The simple dynamo type generator above consists of a permanent magnet which rotates around a central shaft with a coil of wire placed next to this rotating magnetic field. As the magnet spins, the magnetic field around the top and bottom of the coil constantly changes between a north and a south pole. This rotational movement of the magnetic field results in an alternating emf being induced into the coil as defined by Faraday's law of electromagnetic induction.

The magnitude of the electromagnetic induction is directly proportional to the flux density, â the number of loops giving a total length of the conductor, l in meters and the rate or velocity, í at which the magnetic field changes within the conductor in meters/second or m/s, giving by the motional emf expression:

FARADAY'S MOTIONAL EMF EXPRESSION

$$\varepsilon = -\beta.\ell.\upsilon \quad \text{volts}$$

If the conductor does not move at right angles (90°) to the magnetic field then the angle è° will be added to the above expression giving a reduced output as the angle increases:

$$\varepsilon = -\beta.\ell.\upsilon\sin\theta \quad \text{volts}$$

LENZ'S LAW OF ELECTROMAGNETIC INDUCTION

Faraday's Law tells us that inducing a voltage into a conductor can be done by either passing it through a magnetic field, or by moving the magnetic field past the conductor and that if this conductor is part of a closed circuit, an electric current will flow. This voltage is called an induced emf as it has been induced into the conductor by a changing magnetic field due to electromagnetic induction with the negative sign in Faraday's law telling us the direction of the induced current (or polarity of the induced emf).

But a changing magnetic flux produces a varying current through the coil which itself will produce its own magnetic field as we saw in the Electromagnets tutorial. This self-induced emf opposes the change that is causing it and the faster the rate of change of current the greater is the opposing emf. This self-induced emf will, by Lenz's law oppose the change in current in the coil and because of its direction this self-induced emf is generally called a back-emf.

Lenz's Law states that: " the direction of an induced emf is such that it will always opposes the change that is causing it". In other words, an induced current will always OPPOSE the motion or change which started the induced current in the first place and this idea is found in the analysis of Inductance.

Likewise, if the magnetic flux is decreased then the induced emf will oppose this decrease by generating and induced magnetic flux that adds to the original flux.

Lenz's law is one of the basic laws in electromagnetic induction for determining the direction of flow of induced currents and is related to the law of conservation of energy.

According to the law of conservation of energy which states that the total amount of energy in the universe will always remain constant as energy can not be created nor destroyed. Lenz's law is derived from Michael Faraday's law of induction.

One final comment about Lenz's Law regarding electromagnetic induction. We now know that when a relative motion exists between a conductor and a magnetic field, an emf is induced within the conductor.

But the conductor may not actually be part of the coils electrical circuit, but may be the coils iron core or some other metallic part of the system, for example, a transformer. The induced emf within this metallic part of the system causes a circulating current to flow around it and this type of core current is known as an Eddy Current.

Eddy currents generated by electromagnetic induction circulate around the coils core or any connecting metallic components inside the magnetic field because for the magnetic flux they are acting like a single loop of wire. Eddy currents do not contribute anything towards the usefulness of the system but instead they oppose the flow of the induced current by acting like a negative force generating resistive heating and power loss within the core. However, there are electromagnetic induction furnace applications in which only eddy currents are used to heat and melt ferromagnetic metals.

EDDY CURRENTS CIRCULATING IN A TRANSFORMER

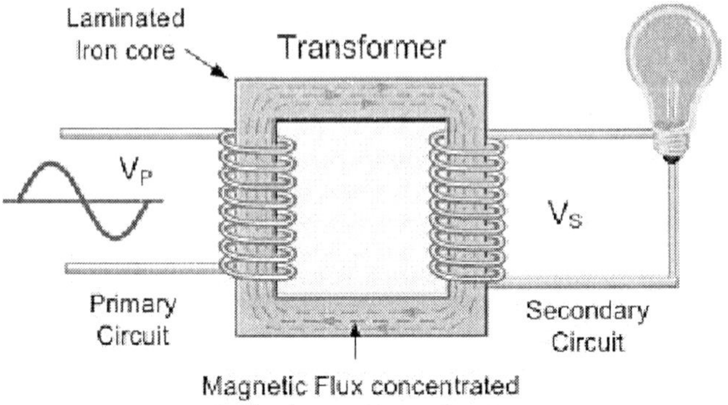

The changing magnetic flux in the iron core of a transformer above will induce an emf, not only in the primary and secondary windings, but also in the iron core. The iron core is a good conductor, so the currents induced in a solid iron core will be large. Furthermore, the eddy currents flow in a direction which, by Lenz's law, acts to weaken the flux created by the primary coil. Consequently, the current in the primary coil required to produce a given B field is increased, so the hysteresis curves are fatter along the H axis.

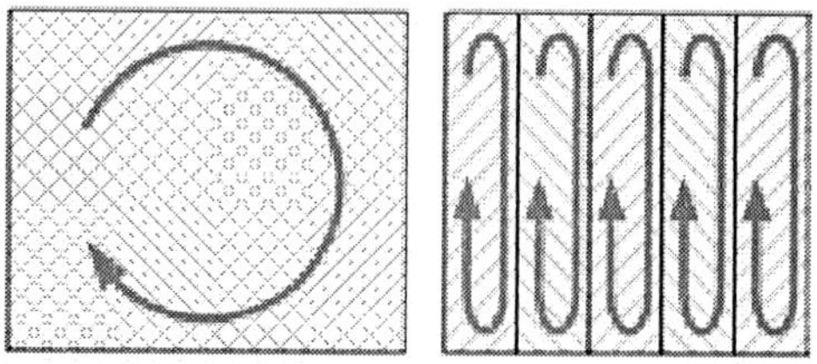

LAMINATING THE IRON CORE

Eddy current and hysteresis losses can not be eliminated completely, but they can be greatly reduced. Instead of having a solid iron core as the magnetic core material of the transformer or coil, the magnetic path is "laminated".

These laminations are very thin strips of insulated (usually with varnish) metal joined together to produce a solid core. The laminations increase the resistance of the iron-core thereby increasing the overall resistance to the flow of the eddy currents, so the induced eddy current power-loss in the core is reduced, and it is for this reason why the magnetic iron circuit of transformers and electrical machines are all laminated.

HOW INDUCTION COOKTOPS WORK

ELECTROMAGNETIC INDUCTION

There's something almost magical about magnetism. As children, we're captivated by a magnet's ability to affect metals like iron, nickel and cobalt without touching them. We learn about attraction and repulsion between magnetic poles and witness the shape of a magnetic field formed in iron filings surrounding a bar magnet. Physicists tell us that electromagnetism, the force governing both electricity and magnetism, is many times stronger than gravity. The suspension of a maglev train above its track is a striking example of this power.

As the name "electromagnetism" suggests, electricity and magnetism are very closely linked. This relationship allows them to affect each other without contact, as in the maglev train example, or through electromagnetic induction. Electromagnetic induction occurs when a circuit with an alternating current flowing through it generates current in another circuit simply by being placed nearby. An alternating current is the kind of electricity flowing through power lines and home wiring, as opposed to a direct current, which we get from batteries.

How does one circuit cause a current in another without touching it, and what does any of this have to do with magnetism? Before we get into that, we need to look at a few principles linking magnetism and electricity:

1. Every electric current has a magnetic field surrounding it.
2. Alternating currents have fluctuating magnetic fields.
3. Fluctuating magnetic fields cause currents to flow in conductors placed within them, which is also known as Faraday's Law.

Adding these three properties together means that a changing electric current is surrounded by an associated changing magnetic field, which in turn generates a changing electrical current in a conductor placed within it, which has its own magnetic field...and so on. It is the electromagnetic equivalent of nesting Matryoshka dolls. Thus, in the case of electromagnetic induction, placing a conductor in the magnetic field surrounding the first current generates the second current.

Induction is the principle that makes electric motors, generators and transformers possible, as well as items closer to home such as rechargeable electric toothbrushes and wireless communication devices. If you own a rice cooker, chances are you already cook using induction. Now let's look at how induced current is used to turn up the heat in induction cooktops.

We have now seen that if electrical current is flowing in a conductor, there is an associated magnetic field created around the wire. In a similar manner, if we move a wire inside a magnetic field there will be an electrical current that will be generated in the wire.

INDUCTION

Current is produced in a conductor when it is moved through a magnetic field because the magnetic lines of force are applying a force on the free electrons in the conductor and causing them to move. This process of generating current in a conductor by placing the conductor in a changing magnetic field is called induction. This is called induction because there is no physical connection between the conductor and the magnet. The current is said to be induced in the conductor by the magnetic field.

One requirement for this electromagnetic induction to take place is that the conductor, which is often a piece of wire, must be perpendicular to the magnetic lines of force in order to produce the maximum force on the free electrons. The direction that the induced current flows is determined by the direction of the lines of force and by the direction the wire is moving in the field. In the animation above the ammeter (the instrument used to measure current) indicates when there is current in the conductor.

If an AC current is fed through a piece of wire, the electromagnetic field that is produced is constantly growing and shrinking due to the constantly changing current in the wire. This growing and shrinking magnetic field can induce electrical current in another wire that is held close to the first wire. The current in the second wire will also be AC and in fact will look very similar to the current flowing in the first wire.

The production of an electromotive force either by motion of a conductor through a magnetic field in such a manner as to cutacross the magnetic flux or by a change in the magnetic flux that threads a conductor

If the flux threading a coil is produced by a current in the coil, any change in that currentwill cause a change in flux, and thusthere will be an induced emf while the current is changing. This process is called self-induction.

The emf of self-induction isproportional to the rate of change of current.

The process by which an emf is induced in one circuit by a change of current in a neighboring circuit is called mutualinduction. Flux produced by a current in a circuit A threads or links circuit B. When there is a change of current in circuit A,there is a change in the flux linking coil B, and an emf is induced in circuit B while the change is taking place. Transformersoperate on the principle of mutual induction.

The phenomenon of electromagnetic induction has a great many important applications in modern technology.

It is common to wrap the wire into a coil to concentrate the strength of the magnetic field at the ends of the coil. Wrapping the coil around an iron bar will further concentrate the magnetic field in the iron bar. The magnetic field will be strongest inside the bar and at its ends (poles).

The production of an electromotive force (induced emf) in a conducting circuit located in a variable magnetic field or moving ina constant magnetic field. The electric current induced by this emf is called an inducedcurrent. Electromagnetic inductionwas discovered by M. Faraday in 1831. According to Faraday's law, theinduced emf Ej in a circuit is directly proportional tothe rate of change of the magnetic flux (the flux of themagnetic induction vector) Õ through the surface S bounded by thiscircuit:

Here ÄÖ is the change in the magnetic flux through the circuit over time Ac the proportionality coefficient k in theInternational System of Units is equal to k = 1 and in the cgs (Gauss) system, k = 1/c, where c is the velocity of light in avacuum. The minus sign defines the direction of the induced current in accordance with Lenz' rule: the direction of theinduced current is such that the magnetic induction flux that it creates through the area bounded by the circuit tends tooppose the flux Ö variation that is responsible for the induced current.

In a constant magnetic field an induced emf arises only during the motion of the circuit in which the magnetic flux through thesurface bounded by the circuit changes with time (that is, the circuit in its motion must intersect the lines of magnetic flux;when the motion is along the lines, the flux Ö will not change and no emf will arise). In this case the induced emf is equal tothe work of the magnetic part of the Lorentz force needed to displace a unit charge along the closed circuit. If, however, afixed conductor is in a constant magnetic field, the induced emf is equal to the work needed to displace a unit charge alongthe closed circuit; this work is performed by the forces of a circuital electric field, which, according to Maxwell's equations,originates in space when a magnetic field changes withtime. In a reference system, with respect to which the circuit is atrest, it is precisely this circuital electric field that induces the motion of charged particles, that is, the appearance of aninduced current. Electromagnetic induction underlies the operation of electric current generators, in which mechanical energyis transformed into electrical; the operation of transformers and other equipment is also based on this phenomenon.

The production of an electromotive force either by motion of a conductor through a magnetic field in such a manner as to cutacross the magnetic flux or by a change in the magnetic flux that threads a conductor. See Electromotive force (emf)

If the flux threading a coil is produced by a current in the coil, any change in that currentwill cause a change in flux, and thusthere will be an induced emf while the current is changing. This process is called self-induction. The emf of self-induction isproportional to the rate of change of current.

The process by which an emf is induced in one circuit by a change of current in a neighboring circuit is called mutualinduction. Flux produced by a current in a circuit A threads or links circuit B. When there is a change of current in circuit A,there is a change in the flux linking coil B, and an emf is induced in circuit B while the change is taking place. Transformersoperate on the principle of mutual induction.

MAXWELL-FARADAY EQUATION

The more generalized equation is one of Maxwell's equations, called the Maxwell-Faraday equation, which defines the relationship between changes in electrical fields and magnetic fields.

It takes the form of:

″×E = – ″B / ″t

where the ″× notation is known as the curl operation, the E is the electric field (a vector quantity) and B is the magnetic field (also a vector quantity). The symbols ″ represent the partial differentials, so the right-hand of the equation is the negative partial differential of the magnetic field with respect to time.

Both E and B are changing in terms of time t, and since they are moving the position of the fields are also changing.

2

Information about Induction

INDUCTION (IN ELECTRICITY AND MAGNETISM)

THE COLUMBIA ENCYCLOPEDIA, 6TH ED.

Induction, in electricity and magnetism, common name for three distinct phenomena. Electromagnetic induction is the production of an electromotive force (emf) in a conductor as a result of a changing magnetic field about the conductor and is the most important of the three phenomena. It was discovered in 1831 by Michael Faraday and independently by Joseph Henry. Variation in the field around a conductor may be produced by relative motion between the conductor and the source of the magnetic field, as in an electric generator, or by varying the strength of the entire field, so that the field around the conductor is also changing. Since a magnetic field is produced around a current-carrying conductor, such a field can be changed by changing the current. Thus, if the conductor in which an emf is to be induced is part of an electric circuit, the induction can be caused by changing the current in that circuit; this is called self-induction. The induced emf is always such that it opposes the change that gives rise to it, according to Lenz's law. Changing the current in a given circuit can also induce an emf in another, nearby circuit unconnected with the original circuit; this type of electromagnetic induction, called mutual induction, is the basis of the transformer. Electrostatic induction is the production of an unbalanced electric charge on an uncharged metallic body as a result of a charged body being brought near it without touching it. If the charged body is positively charged, electrons in the uncharged body will be attracted toward it; if the opposite end of the body is then grounded, electrons will flow onto it to replace those drawn to the other end, the body thus acquiring a negative charge after the ground connection is broken. A similar procedure can be used to produce a positive charge on the uncharged body when a negatively charged body is brought near it. See electricity. Magnetic induction is the production of a magnetic field in a piece of unmagnetized iron or other ferromagnetic substance when a magnet is brought near it. The magnet causes the individual particles of the iron, which act like tiny magnets, to line up so that the sample as a whole becomes magnetized. Most of this induced magnetism is lost when the magnet causing it is taken away.

Also Known As: induction (not to be confused with inductive reasoning), Faraday's law of electromagnetic induction.

Magnetism,force of attraction or repulsion between various substances, especially those made of iron and certain other metals; ultimately it is due to the motion of electric charges.

MAGNETIC POLES, FORCES, AND FIELDS

Any object that exhibits magnetic properties is called a magnet. Every magnet has two points, or poles, where most of its strength is concentrated; these are designated as a north-seeking pole, or north pole, and a south-seeking pole, or south pole, because a suspended magnet tends to orient itself along a north-south line. Since a magnet has two poles, it is sometimes called a magnetic dipole, being analogous to an electric dipole, composed of two opposite charges. The like poles of different magnets repel each other, and the unlike poles attract each other.

One remarkable property of magnets is that whenever a magnet is broken, a north pole will appear at one of the broken faces and a south pole at the other, such that each piece has its own north and south poles. It is impossible to isolate a single magnetic pole, regardless of how many times a magnet is broken or how small the fragments become. (The theoretical question as to the possible existence in any state of a single magnetic pole, called a monopole, is still considered open by physicists; experiments to date have failed to detect one.)

From his study of magnetism, C. A. Coulomb in the 18th cent.found that the magnetic forces between two poles followed an inverse-square law of the same form as that describing the forces between electric charges. The law states that the force of attraction or repulsion between two magnetic poles is directly proportional to the product of the strengths of the poles and inversely proportional to the square of the distance between them.

As with electric charges, the effect of this magnetic force acting at a distance is expressed in terms of a field of force. A magnetic pole sets up a field in the space around it that exerts a force on magnetic materials. The field can be visualized in terms of lines of induction (similar to the lines of force of an electric field). These imaginary lines indicate the direction of the field in a given region. By convention they originate at the north pole of a magnet and form loops that end at the south pole either of the same magnet or of some other nearby magnet (see also flux, magnetic). The lines are spaced so that the number per unit area is proportional to the field strength in a given area. Thus, the lines converge near the poles, where the field is strong, and spread out as their distance from the poles increases.

A picture of these lines of induction can be made by sprinkling iron filings on a piece of paper placed over a magnet. The individual pieces of iron become magnetized by entering a magnetic field, i.e., they act like tiny magnets, lining themselves up along the lines of induction. By using variously shaped magnets and various combinations of more than one magnet, representations of the field in these different situations can be obtained.

MAGNETIC MATERIALS

The term magnetism is derived from Magnesia, the name of a region in Asia Minor where lodestone, a naturally magnetic iron ore, was found in ancient times. Iron is not the only material that is easily magnetized when placed in a magnetic field; others include nickel and cobalt. Carbon steel was long the material commonly used for permanent magnets, but more recently other materials have been developed that are much more efficient as permanent magnets, including certain ferroceramics and Alnico, an alloy containing iron, aluminum, nickel, cobalt, and copper.

Materials that respond strongly to a magnetic field are called ferromagnetic [Lat. ferrum=iron]. The ability of a material to be magnetized or to strengthen the magnetic field in its vicinity is expressed by its magnetic permeability. Ferromagnetic materials have permeabilities of as much as 1,000 or more times that of free space (a vacuum). A number of materials are very weakly attracted by a magnetic field, having permeabilities slightly greater than that of free space; these materials are called paramagnetic. A few materials, such as bismuth and antimony, are repelled by a magnetic field, having permeabilities less than that of free space; these materials are called diamagnetic.

THE BASIS OF MAGNETISM

The electrical basis for the magnetic properties of matter has been verified down to the atomic level. Because the electron has both an electric charge and a spin, it can be called a charge in motion. This charge in motion gives rise to a tiny magnetic field. In the case of many atoms, all the electrons are paired within energy levels, according to the exclusion principle, so that the electrons in each pair have opposite (antiparallel) spins and their magnetic fields cancel. In some atoms, however, there are more electrons with spins in one direction than in the other, resulting in a net magnetic field for the atom as a whole; this situation exists in a paramagnetic substance. If such a material is placed in an external field, e.g., the field created by an electromagnet, the individual atoms will tend to align their fields with the external one. The alignment will not be complete, due to the disruptive effect of thermal vibrations. Because of this, a paramagnetic substance is only weakly attracted by a magnet.

In a ferromagnetic substance, there are also more electrons with spins in one direction than in the other. The individual magnetic fields of the atoms in a given region tend to line up in the same direction, so that they reinforce one another. Such a region is called a domain. In an unmagnetized sample, the domains are of different sizes and have different orientations. When an external magnetic field is applied, domains whose orientations are in the same general direction as the external field will grow at the expense of domains with other orientations. When the domains in all other directions have vanished, the remaining domains are rotated so that their direction is exactly the same as that of the external field. After this rotation is complete, no further magnetization can take place, no matter how strong the external field; a saturation point is said to have been reached. If the external field is then reduced to zero, it is found that the sample still retains some of its magnetism; this is known as hysteresis.

EVOLUTION OF ELECTROMAGNETIC THEORY

The connections between magnetism and electricity were discovered in the early part of the 19th cent. In 1820 H. C. Oersted found that a wire carrying an electrical current deflects the needle of a magnetic compass because a magnetic field is created by the moving electric charges constituting the current. It was found that the lines of induction of the magnetic field surrounding the wire (or any other conductor) are circular. If the wire is bent into a coil, called a solenoid, the magnetic fields of the individual loops combine to produce a strong field through the core of the coil. This field can be increased manyfold by inserting a piece of soft iron or other ferromagnetic material into the core; the resulting arrangement constitutes an electromagnet.

Following Oersted's discovery the various magnetic effects of an electric current were extensively investigated by J. B. Biot, Félix Savart, and A. M. Ampère. Ampère showed in 1825 that not only does a current-carrying conductor exert a force on a magnet but magnets also exert forces on current-carrying conductors. In 1831 Michael Faraday and Joseph Henry independently discovered that it is possible to produce a current in a conductor by changing the magnetic field about it. The discovery of this effect, called electromagnetic induction, together with the discovery that an electric current produces a magnetic field, laid the foundation for the modern age of electricity. Both the electric generator, which makes electricity widely available, and the electric motor, which converts electricity to useful mechanical work, are based on these effects.

Another relationship between electricity and magnetism is that a regularly changing electric current in a conductor will create a changing magnetic field in the space about the conductor, which in turn gives rise to a changing electrical field. In this way regularly oscillating electric and magnetic fields can generate each other. These fields can be visualized as a single wave that is propagating through space. The formal theory underlying this electromagnetic radiation was developed by James Clerk Maxwell in the middle of the 19th cent. Maxwell showed that the speed of propagation of electromagnetic radiation is identical with that of light, thus revealing that light is intimately connected with electricity and magnetism.

MAGNETISM

The branch of science that describes the effects of the interactions between charges due to their motion and spin. Theseinteractions may appear in various forms, including electric currents and permanent magnets. They are described in terms ofthe magnetic field, although the field hypothesis cannot be tested independently of the electrokinetic effects by which it isdefined. The magnetic field complements the concept of the electrostatic field used to describe the potential energy betweencharges due to their relative positions. Special relativity theory relates the two, showing that magnetism is a relativisticmodification of the electrostatic forces. The two together form the electromagnetic interactions which are propagated aselectromagnetic waves, including light. They control the structure of materials at distances between the long-rangegravitational actions and the short-range "strong" and "weak" forces most evident within the atomic nucleus. SeeElectromagnetic radiation, Relativity

The magnetic field can be visualized as a set of lines (Fig. 1) illustrated by iron filings scattered on a suitable surface. Theintensity of the field is indicated by the line spacing, and the direction by arrows pointing along the lines. The sign conventionis chosen so that the Earth's magnetic field is directed from the north magnetic pole toward the south magnetic pole. Thefield can be defined and measured in various ways, including the forces on the equivalent magnetic poles, and on currents ormoving charges. Bringing a coil of wire into the field, or removing it, induces an electromotive force (emf) which depends onthe rate at which the number of field lines, referred to as lines of magnetic flux, linking the coil changes in time. This providesa definition of flux, &PHgr;, in terms of the emf, e, given by Eq. (1)

(1) $e = -N\, d\Phi/dt$ volts

for a coil of N turns wound sufficiently closely to make the number of lines linking each the same. The International System(SI) unit of &PHgr;, the weber (Wb), is defined accordingly as the volt-second. The symbol B is used to denote the flux, orline, density, as in Eq. (2),

(2) $B = \Phi/\text{area}$

when the area of the coil is sufficiently small to sample conditions at a point, and the coil is oriented so that the induced emfis a maximum. The SI unit of B, the tesla (T), is the Wb/m. The sign of the emf, e, is measured positively in the direction of aright-hand screw pointing in the direction of the flux lines. It is often convenient, particularly when calculating induced emfs, todescribe the field in terms of a magnetic vector potential function instead of flux.

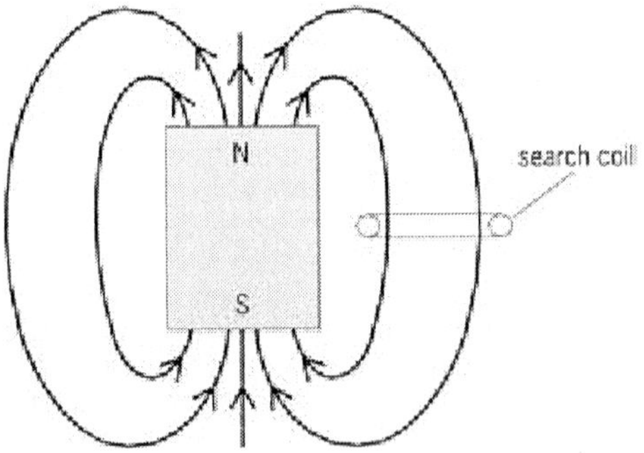

Magnetic lines of a bar magnet

MAGNETIC CIRCUITS

The magnetic circuit provides a useful method of analyzing devices with ferromagnetic parts, and introduces variousquantities used in magnetism. It describes the use of ferromagnetic materials to control the flux paths in a manner analogousto the role of conductors in carrying currents around electrical circuits. For example, pieces of iron may be used to guide theflux which is produced by a magnet along a path which includes an air gap (Fig. 2), giving an increase in the flux density, B, ifthe cross-sectional area of the gap is less than that of the magnet. See Magnet, Magnetic materials

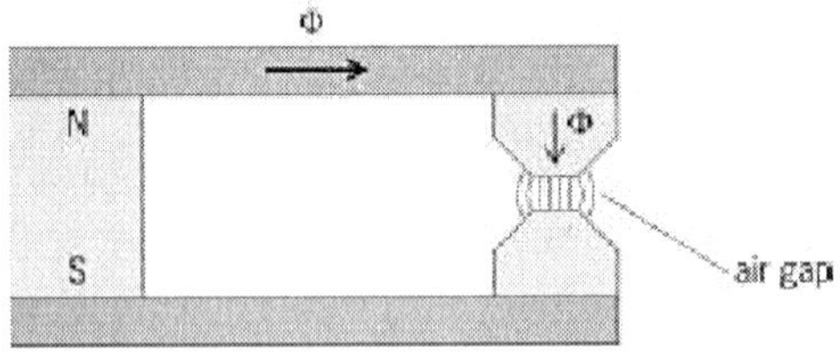

Magnetic circuit with an air gap

The magnet may be replaced by a coil of N turns carrying a current, i, wound over a piece of iron, or ferromagnetic material,in the form of a ring of uniform cross section. The flux linking each turn of the coil, and each turn of a secondary coil woundseparately from the first, is then approximately the same, giving the same induced emf per turn [according to Eq. (1)] whenthe supply current, i, and hence the flux, &PHgr;, changes in time. The arrangement is typical of many different devices. Itprovides, for example, an electrical transformer whose input and output voltages are directly proportional to the numbers ofturns in the windings. Emf's also appear within the iron, and tend to produce circulating currents and losses. These arecommonly reduced by dividing the material into thin laminations. See Eddy current

The amount of flux produced by a given supply current is reduced by the presence of any air gaps which may be introducedto contribute constructional convenience or to allow a part to move. The effects of the gaps, and of different magneticmaterials, can be predicted by utilizing the analogy between flux, &PHgr;, and the flow of electric current through a circuitconsisting of resistors connected in series (Fig. 3). Since &PHgr; depends on the product, iN, of the winding current andnumber

(3) $iN = \Phi\mathfrak{R}$

of turns, as in Eq. (3), the ratio between them, termed the reluctance,, is the analog of electrical resistance. It may beconstant or may vary with &PHgr;. The quantity iN is the magnetomotive force (mmf), analogous to voltage or emf in theequivalent electrical circuit. The relationship between the two exchanges the potental and flow quantities, since the magneticmmf depends on current, i, and the electrical emf on d&PHgr;/dt. Electric and magnetic equivalent circuits are referred to asduals. See Reluctance

MAGNETISM

CONCEPT

Most people are familiar with magnets primarily as toys, or as simple objects for keeping papers attached to a metal surface such as a refrigerator door. In fact the areas of application for magnetism are much broader, and range from security to health care to communication, transportation, and numerous other aspects of daily life. Closely related to electricity, magnetism results from specific forms of alignment on the part of electron charges in certain varieties of metal and alloy.

HOW IT WORKS

Magnetism, along with electricity, belongs to a larger phenomenon, electromagnetism, or the force generated by the passage of an electric current through matter. When two electric charges are at rest, it appears to the observer that the force between them is merely electric. If the charges are in motion, however—and in this instance motion or rest is understood in relation to the observer—then it appears as though a different sort of force, known as magnetism, exists between them.

REPORT ADVERTISEMENT

In fact, the difference between magnetism and electricity is purely artificial. Both are manifestations of a single fundamental force, with "magnetism" simply being an abstraction that people use for the changes in electromagnetic force created by the motion of electric charges. It is a distinction on the order of that between water and wetness; nonetheless, it is often useful and convenient to discuss the two phenomena as though they were separate.

At the atomic level, magnetism is the result of motion by electrons, negatively charged subatomic particles, relative to one another. Rather like planets in a solar system, electrons both revolve around the atom's nucleus and rotate on their own axes. (In fact the exact nature of their movement is much more complex, but this analogy is accurate enough for the present purposes.) Both types of movement create a magnetic force field between electrons, and as a result the electron takes on the properties of a tiny bar magnet with a north pole and south pole. Surrounding this infinitesimal magnet are lines of magnetic force, which begin at the north pole and curve outward, describing an ellipse as they return to the south pole.

In most atomic elements, the structure of the atom is such that the electrons align in a random manner, rather like a bunch of basketballs bumping into one another as they float in a swimming pool. Because of this random alignment, the small magnetic fields cancel out one another. Two such self-canceling particles are referred to as paired electrons, and again, the analogy to bar magnets is an appropriate one: if one were to shake a bag containing an even number of bar magnets, they would all wind up in pairs, joined at opposing (north-south) poles.

There are, however, a very few elements in which the fields line up to create what is known as a net magnetic dipole, or a unity of direction—rather like a bunch of basketballs simultaneously thrown from in the same direction at the same time. These elements, among them iron, cobalt, and nickel, as well as various alloys or mixtures, are commonly known as magnetic metals or natural magnets.

It should be noted that in magnetic metals, magnetism comes purely from the alignment of forces exerted by electrons as they spin on their axes, whereas the forces created by their orbital motion around the nucleus tend to cancel one another out. But in magnetic rare earth elements such as cerium, magnetism comes both from rotational and orbital forms of motion. Of principal concern in this discussion, however, is the behavior of natural magnets on the one hand, and of nonmagnetic materials on the other.

There are five different types of magnetism—diamagnetism, paramagnetism, ferromagnetism, ferrimagnetism, and antiferromagnetism. Actually, these terms describe five different types of response to the process of magnetization, which occurs when an object is placed in a magnetic field.

A magnetic field is an area in which a magnetic force acts on a moving charged particle such that the particle would experience no force if it moved in the direction of the magnetic field—in other words, it would be "drawn," as a ten-penny nail is drawn to a common bar or horseshoe (U-shaped) magnet. An electric current is an example of a moving charge, and indeed one of the best ways to create a magnetic field is with a current. Often this is done by means of a solenoid, a current-carrying wire coil through which the material to be magnetized is passed, much as one would pass an object through the interior of a spring.

All materials respond to a magnetic field; they just respond in different ways. Some non-magnetic substances, when placed within a magnetic field, slightly reduce the strength of that field, a phenomenon known as diamagnetism. On the other hand, there are nonmagnetic substances possessing an uneven number of electrons per atom, and in these instances a slight increase in magnetism, known as paramagnetism, occurs. Paramagnetism always has to overcome diamagnetism, however, and hence the gain in magnetic force is very small. In addition, the thermal motion of atoms and molecules prevents the objects' magnetic fields from coming into alignment with the external field. Lower temperatures, on the other hand, enhance the process of paramagnetism.

In contrast to diamagnetism and paramagnetism, ferro-, ferri-, and antiferromagnetism all describe the behavior of natural magnets when exposed to a magnetic field. The name ferromagnetism suggests a connection with iron, but in fact the term can apply to any of those materials in which the magnitude of the object's magnetic field increases greatly when it is placed within an external field. When a natural magnet becomes magnetized (that is, when a metal or alloy comes into contact with an external magnetic field), a change occurs at the level of the domain, a group of atoms equal in size to about 5 ×10 meters across—just large enough to be visible under a microscope.

In an unmagnetized sample, there may be an alignment of unpaired electron spins within a domain, but the direction of the various domains' magnetic forces in relation to one another is random. Once a natural magnet is placed within an external magnetic force field, however, one of two things happens to the domains. Either they all come into alignment with the field or, in certain types of material, those domains in alignment with the field grow while the others shrink to nonexistence.

The first of these processes is called domain alignment or ferromagnetism, the second domain growth or ferrimagnetism. Both processes turn a natural magnet into what is known as a permanent magnet—or, in common parlance, simply a "magnet." The latter is then capable of temporarily magnetizing a ferromagnetic item, as for instance when one rubs a paper clip against a permanent magnet and then uses the magnetized clip to lift other paper clips. Of the two varieties, however, a ferromagnetic metal is stronger because it requires a more powerful magnetic force field in order to become magnetized. Most powerful of all is a saturated ferromagnetic metal, one in which all the unpaired electron spins are aligned.

Once magnetized, it is very hard for a ferro-magnetic metal to experience demagnetization, or antiferromagnetism. Again, there is a connection between temperature and magnetism, with heat acting as a force to reduce the strength of a magnetic field. Thus at temperatures above 1,418°F (770°C), the atoms within a domain take on enough kinetic energy to overpower the forces holding the electron spins in alignment. In addition, mechanical disturbances—for instance, battering a permanent magnet with a hammer—can result in some reduction of magnetic force.

Many of the best permanent magnets are made of steel, which, because it is an alloy of iron with carbon and other elements, has an irregular structure that lends itself well to the ferromagnetic process of domain alignment. Iron, by contrast, will typically lose its magnetization when an external magnetic force field is removed; but this actually makes it a better material for some varieties of electromagnet.

The latter, in its simplest form, consists of an iron rod inside a solenoid. When a current is passing through the solenoid, it creates a magnetic force field, activating the iron rod and turning it into an electromagnet. But as soon as the current is turned off, the rod loses its magnetic force. Not only can an electromagnet thus be controlled, but it is often stronger than a permanent magnet: hence, for instance, giant electromagnets are used for lifting cars in junkyards.

REAL-LIFE APPLICATIONS

FINDING THE WAY: MAGNETS IN COMPASSES

A north-south bar magnet exerts exactly the same sort of magnetic field as a solenoid. Lines of magnetic run through it in one direction, from "south" to "north," and upon leaving the north pole of the magnet, these lines describe an ellipse as they curve back around to the south pole. In view of this model, it is also easy to comprehend why a pair of opposing poles attracts one another, and a pair of like poles—for whom the lines of force are moving away from each other—repels. This is a fact particularly applicable to the operation of MAGLEV trains, as discussed later.

A magnetic compass works because Earth itself is like a giant bar magnet, complete with vast arcs of magnetic force, called the geomagnetic field, surrounding the planet. The first scientist to recognize the magnetic properties of Earth was the English physicist William Gilbert (1544-1603). Scientists today believe that the source of Earth's magnetism lies in a core of molten iron some 4,320 mi (6,940 km) across, constituting half the planet's diameter. Within this core run powerful electric currents that ultimately create the geomagnetic field.

Just as a powerful magnet causes all the domains in a magnetic metal to align with it, a bar magnet placed in a magnetic field will rotate until it lines up with the field's direction. The same thing happens when one suspends a magnet from a string: it lines up with Earth's magnetic field, and points in a north-south direction. The Chinese of the first century b.c., though unaware of the electromagnetic forces that caused this to happen, discovered that a strip of magnetic metal always tended to point toward geographic north.

This led ultimately to the development of the magnetic compass, which typically consists of a magnetized iron needle suspended over a card marked with the four cardinal directions. The needle is attached to a pivoting mechanism at its center, which allows it to move freely so that the "north" end will always point the user northward.

The magnetic compass proved so important that it is typically ranked alongside paper, printing, and gunpowder as one of premodern China's four great gifts to the West. Prior to the compass, mariners had to depend purely on the position of the Sun and other, less reliable, means of determining direction; hence the invention quite literally helped open up the world. But there is a somewhat irksome anomaly lurking in the seeming simplicity of the magnetic compass.

In fact magnetic north is not the same thing as true north; or, to put it another way, if one continued to follow a compass northward, it would lead not to the Earth's North Pole, but to a point identified in 1984 as 77°N, 102°182 W—that is, in the Queen Elizabeth Islands of far northern Canada. The reason for this is that Earth's magnetic field describes a current loop whose center is 11° off the planet's equator, and thus the north and south magnetic poles—which are on a plane perpendicular to that of the Earth's magnetic field—are 11° off of the planet's axis.

REPORT ADVERTISEMENT

The magnetic field of Earth is changing position slowly, and every few years the United States Geological Survey updates magnetic declination, or the shift in the magnetic field. In addition, Earth's magnetic field is slowly weakening as well. The behavior, both in terms of weakening and movement, appears to be similar to changes taking place in the magnetic field of the Sun.

MAGNETS FOR DETECTION: BURGLAR ALARMS, MAGNETOMETERS, AND MRI

A compass is a simple magnetic instrument, and a burglar alarm is not much more complex. A magnetometer, on the other hand, is a much more sophisticated piece of machinery for detecting the strength of magnetic fields. Nonetheless, the magnetometer bears a relation to its simpler cousins: like a compass, certain kinds of magnetometers respond to a planet's magnetic field; and like a burglar alarm, other varieties of magnetometer are employed for security.

At heart, a burglar alarm consists of a contact switch, which responds to changes in the environment and sends a signal to a noisemaking device. The contact switch may be mechanical—a simple fastener, for instance—or magnetic. In the latter case, a permanent magnet may be installed in the frame of a window or door, and a piece of magnetized material in the window or door itself. Once the alarm is activated, it will respond to any change in the magnetic field—i.e., when someone slides open the door or window, thus breaking the connection between magnet and metal.

Though burglar alarms may vary in complexity, and indeed there may be much more advanced systems using microwaves or infrared rays, the application of magnetism in home security is a simple matter of responding to changes in a magnetic field. In this regard, the principle governing magnetometers used at security checkpoints is even simpler. Whether at an airport or at the entrance to some other high-security venue, whether handheld or stationary, a magnetometer merely detects the presence of magnetic metals. Since the vast majority of firearms, knife-blades, and other weapons are made of iron or steel, this provides a fairly efficient means of detection.

At a much larger scale, magnetometers used by astronomers detect the strength and sometimes the direction of magnetic fields surrounding Earth and other bodies in space. This variety of magnetometer dates back to 1832, when mathematician and scientist Carl Friedrich Gauss (1777-1855) developed a simple instrument consisting of a permanent bar magnet suspended horizontally by means of a gold wire. By measuring the period of the magnet's oscillation in Earth's magnetic field (or magnetosphere), Gauss was able to measure the strength of that field. Gauss's name, incidentally, would later be applied to the term for a unit of magnetic force. The gauss, however, has in recent years been largely replaced by the tesla, named after Nikola Tesla (1856-1943), which is equal to one newton/ampere meter (1 N/A·m) or 10 (10,000) gauss.

As for magnetometers used in astronomical research, perhaps the most prominent—and certainly one of the most distant—ones is on Galileo, a craft launched by the U.S. National Aeronautics and Space Administration (NASA) toward Jupiter on October 15, 1989. Among other instruments on board Galileo, which has been in orbit around the solar system's largest planet since 1995, is a magnetometer for measuring Jupiter's magnetosphere and that of its surrounding asteroids and moons.

Closer to home, but no less impressive, is another application of magnetism for the purposes of detection: magnetic resonance imagining, or MRI. First developed in the early 1970s, MRI permits doctors to make intensive diagnoses without invading the patient's body either with a surgical knife or x rays.

The heart of the MRI machine is a large tube into which the patient is placed in a supine position. A technician then activates a powerful magnetic field, which causes atoms within the patient's body to spin at precise frequencies. The machine then beams radio signals at a frequency matching that of the atoms in the cells (e.g., cancer cells) being sought. Upon shutting off the radio signals and magnetic field, those atoms emit bursts of energy that they have absorbed from the radio waves. At that point a computer scans the body for frequencies matching specific types of atoms, and translates these into three-dimensional images for diagnosis.

Magnets for Projecting Sound: Microphones, Loudspeakers, Car Horns, and Electric Bells

The magnets used in Galileo or an MRI machine are, needless to say, very powerful ones, and as noted earlier, the best way to create a super-strong, controllable magnet is with an electrical current. When that current is properly coiled around a magnetic metal, this creates an electromagnet, which can be used in a variety of applications.

As discussed above, the most powerful electromagnets typically use nonpermanent magnets so as to facilitate an easy transition from an extremely strong magnetic field to a weak or nonexistent one. On the other hand, permanent magnets are also used in loudspeakers and similar electromagnetic devices, which seldom require enormous levels of power.

In discussing the operation of a loudspeaker, it is first necessary to gain a basic understanding of how a microphone works. The latter contains a capacitor, a system for storing charges in the form of an electrical field. The capacitor's negatively charged plate constitutes the microphone's diaphragm, which, when it is hit by sound waves, vibrates at the same frequency as those waves. Current flows back and forth between the diaphragm and the positive plate of the capacitor, depending on whether the electrostatic or electrical pull is increasing or decreasing. This in turn produces an alternating current, at the same frequency as the sound waves, which travels through a mixer and then an amplifier to the speaker.

A loudspeaker typically contains a circular permanent magnet, which surrounds an electrical coil and is in turn attached to a cone-shaped diaphragm. Current enters the speaker ultimately from the microphone, alternating at the same frequency as the source of the sound (a singer's voice, for instance). As it enters the coil, tnis current induces an alternating magnetic field, which causes the coil to vibrate. This in turn vibrates the cone-shaped diaphragm, and the latter reproduces sounds generated at the source.

A car horn also uses magnetism to create sound by means of vibration. When a person presses down on the horn embedded in his or her steering wheel, this in turn depresses an iron bar that passes through an electromagnet surrounded by wires from the car's battery. The bar moves up and down within the electromagnetic field, causing the diaphragm to vibrate and producing a sound that is magnified greatly when released through a bell-shaped horn.

Electromagnetically induced vibration is also the secret behind another noise-making device, a vibrating electric doorbell used in many apartments. The button that a visitor presses is connected directly to a power source, which sends current flowing through a spring surrounding an electromagnet. The latter generates a magnetic field, drawing toward it an iron armature attached to a hammer. The hammer then strikes the bell. The result is a mechanical reaction that pushes the armature away from the electromagnet, but the spring forces the armature back against the electromagnet again. This cycle of contact and release continues for as long as the button is depressed, causing a continual ringing of the bell.

Recording and Reading Data Using Magnets: From Records and Tapes to Disk Drives

Just as magnetism plays a critical role in projecting the volume of sound, it is also crucial to the recording and retrieval of sound and other data. Of course terms such as "retrieval" and "data" have an information-age sound to them, but the idea of using magnetism to record sound is an old one—much older than computers or compact discs (CDs). The latter, of course, replaced cassettes in the late 1980s as the preferred mode for listening to recorded music, just as cassettes had recently made powerful gains against phonograph records.

Despite the fact that cassettes entered the market much later than records, however, recording engineers from the mid-twentieth century onward typically used magnetic tape for master recordings of songs. This master would then be used to create a metal master record disk by means of a cutting head that responded to vibrations from the master tape; then, the record company could produce endless plastic copies of the metal record.

In recording a tape—whether a stereo master or a mere home recording of a conversation—the principles at work are more or less the same. As noted in the earlier illustration involving a microphone and loudspeaker, sound comes through a microphone in the form of alternating current. The strength of this current in turn affects the "recording head," a small electromagnet whose magnetic field extends over the section of tape being recorded. Loud sounds produce strong magnetic fields, and soft ones weak fields.

All of this information becomes embedded on the cassette tape through a process of magnetic alignment not so different from the process described earlier for creating a permanent magnet. But whereas the permanent magnetization of a natural magnet is difficult to reverse, reversal of a tape's magnetization—in other words, erasing the tape—is easy. An erase head, an electromagnet operating at a frequency too high for the human ear to hear, simply scrambles the magnetic particles on a piece of tape.

A CD, as one might expect, is much more closely related to a computer disk-drive than it is to earlier forms of recording technology. The disk drive receives electronic on-off signals from the computer, and translates these into magnetic codes that it records on the surface of a floppy disk. The disk drive itself includes two electric motors: a disk motor, which rotates the disk at a high speed, and a head motor, which moves the computer's read-write head across the disk. (It should be noted that most electric motors, including the universal motors used in a variety of household appliances, also use electromagnets.)

A third motor, called a stepper motor, ensures that the drive turns at a precise rate of speed. The stepper motor contains its own magnet, in this case a permanent one of cylindrical shape that sends signals to rows of metal teeth surrounding it, and these teeth act as gears to regulate the drive's speed. Likewise a CD player, which actually uses laser beams rather than magnetic fields to retrieve data from a disc, also has a drive system that regulates the speed at which the disk spins.

MAGLEV TRAINS: THE FUTURE OF TRANSPORT?

One promising application of electromagnetic technology relates to a form of transportation that might, at first glance, appear to be old news: trains. But MAGLEV, or magnetic levitation, trains are as far removed from the old steam engines of the Union Pacific as the space shuttle is from the Wright brothers' experimental airplane.

As discussed earlier, magnetic poles of like direction (i.e., north-north or south-south) repel one another such that, theoretically at least, it is possible to keep one magnet suspended in the air over another magnet. Actually it is impossible to produce these results with simple bar magnets, because their magnetic force is too small; but an electromagnet can create a magnetic field powerful enough that, if used properly, it exerts enough repulsive force to lift extremely heavy objects. Specifically, if one could activate train tracks with a strong electromagnetic field, it might be possible to "levitate" an entire train. This in turn would make possible a form of transport that could move large numbers of people in relative comfort, thus decreasing the environmental impact of automobiles, and do so at much higher speeds than a car could safely attain.

Actually the idea of MAGLEV trains goes back to a time when trains held complete supremacy over automobiles as a mode of transportation: specifically, 1907, when rocket pioneer Robert Goddard (1882-1945) wrote a story describing a vehicle that traveled by means of magnetic levitation. Just five years later, French engineer Emile Bachelet produced a working model for a MAGLEV train. But the amount of magnetic force required to lift such a vehicle made it impractical, and the idea fell to the wayside.

Then, in the 1960s, the advent of superconductivity—the use of extremely low temperatures, which facilitate the transfer of electrical current through a conducting material with virtually no resistance—made possible electromagnets of staggering force. Researchers began building MAGLEV prototypes using superconducting coils with strong currents to create a powerful magnetic field. The field in turn created a repulsive force capable of lifting a train several inches above a railroad track. Electrical current sent through guideway coils on the track allowed for enormous propulsive force, pushing trains forward at speeds up to and beyond 250 MPH (402 km/h).

Initially, researchers in the United States were optimistic about MAGLEV trains, but safety concerns led to the shelving of the idea for several decades. Meanwhile, other industrialized nations moved forward with MAGLEVs: in Japan, engineers built a 27-mi (43.5-km) experimental MAGLEV line, while German designers experimented with attractive (as opposed to repulsive) force in their Transrapid 07. MAGLEV trains gained a new defender in the United States with now-retired Senator Daniel Patrick Moynihan (D-NY), who as chairman of a Senate subcommittee overseeing the interstate highway system introduced legislation to fund MAGLEV research. The 1998 transportation bill allocated $950 million toward the Magnetic Levitation Prototype Development Program. As part of this program, in January 2001 the U.S. Department of Transportation selected projects in Maryland and Pennsylvania as the two finalists in the competition to build the first MAGLEV train service in the United States. The goal is to have the service in place by approximately 2010.

DEFINITION OF MAGNETIC FIELD

ELECTROMAGNET:

A type of magnet in which an object is charged by an electrical current. Typically the object used is made of iron, which quickly loses magnetic force when current is reduced. Thus an electromagnet can be turned on or off, and its magnetic force altered, making it potentially much more powerful than a natural magnet.

ELECTROMAGNETISM:

The unified electrical and magnetic force field generated by the passage of an electric current through matter.

ELECTRONS:

Negatively charged subatomic particles whose motion relative to one another creates magnetic force.

MAGNETIC FIELD:

Wherever a magnetic force acts on a moving charged particle, a magnetic field is said to exist. Magnetic fields are typically measured by a unitcalled a tesla.

NATURAL MAGNET:

A chemical element in which the magnetic fields created by electrons' relative motion align uniformly to create a net magnetic dipole, or unity of direction. Such elements, among them iron, cobalt, and nickel, are also known as magnetic metals.

PERMANENT MAGNET:

A magnetic material in which groups of atoms, known as domains, are brought into alignment, and in which magnetization cannot be changed merely by attempting to realign the domains. Permanent magnetization is reversible only at very high temperatures—for example, 1,418°F (770°C) in the case of iron.

INDUCTION COOKER- THE HISTORY BEHIND

Induction cooking is not a new technology, it has long been widely used around the world. People witnessed magic of induction cooking first time in early 1900s. Induction cooking technology was introduced at Chicago in a 'World Fair" in 1933 as well in the mid-1950s. Induction cooker demonstrations were held by GM in North America by the Frigidaire division of GM. To demonstrate convenience and safety of Induction cooking the induction cooker demonstration was shown by placing a newspaper between Induction cooker's surface and the pot while boiling a pot of water. But it never quite caught on and induction cooker productions were delayed for another few more years and for subsequent 40 years the technology was used mostly in industrial applications.

Modern developments and implementations of Induction cooking starts in early 1970s at the R&D Center of Westinghouse Electric Corporation in USA. These modern Induction cooktops used transistors developed for automotive electronic ignition systems to drive the 25 kHz current. The standalone single-burner range was named the Cool Top Induction Range.

Westinghouse developed their Cool Top 2 (CT2) Induction ranges by a team led by Bill Moreland and Terry Malarkey. Their induction ranges included a set of high quality cookware made of Quadruply and were priced at US$1500. Production took place in 1973 through 1975, and stopped coincidentally with the sale of Westinghouse Consumer Products Division to White Consolidated Industries Inc.

In the US, NASA developed it for the space program. Consumer units followed in the US, but were plagued by low power, reliability and noise problems.

Though induction cookers faded from the American consumer market, it continued to be developed in Europe and Asia where energy availability is an issue. Some US manufacturers like Cooktek and Luxine continued to develop commercial units. In 2000, European manufacturers made a breakthrough (in conjunction with DuPont) in insulating materials design for integrating the electronics with the induction generator coils.

The snap together design of these fourth generation systems along with the reduced fabrication costs enabled the manufacturers to produce induction generators for far less than previously, with much more compact designs that were inherently more reliable. As a result the market in Europe really took off. Currently induction cooktops are a norm in new construction in many European countries and are only about 20–30% more expensive than radiant ceramic cooktops there. In Asia a similar phenomenon has occurred. Huge numbers of Asian households are switching to induction for their cooking due to the safer and cooler cooking environment it provides.

With recent improvements in technology, induction cookers are now better than ever and while cheaper manufacturing from China has reduced the cost of induction cookers to levels that are being affordable to every household.

INDUCTION COOKERS - TECHNOLOGY

LET'S LEARN ABOUT INDUCTION COOKING TECHNOLOGY

As we know that the latest cooking technology in the world is induction cooking. The technology behind is quite simple.The basic concept is that induction cooker works on electromagnetic induction principle.The electromagnetic induction principle can take place only in steel and iron utensil. The induction coil inside the induction cooker generates magnetic field and eddy current in the steel or iron utensils.due to this the iron molecules starts rapid movement and give resistance to each other, Due to this heat is generated in the utensils very quickly.

In somany cooking technology, the best available technology is induction cooking,Induction cooking completely different from all other cooking technologies it does not involve generating heat which is then transferred to the cooking vessel, it makes the cooking vessel itself the original generator of the cooking heat.

HOW DOES AN INDUCTION COOKER DO THAT?

Put simply, an induction-cooker element/induction coil (what on a gas stove would be called a "burner") is a powerful, high-frequency electromagnet, with the electromagnetism generated by sophisticated electronics in the "element" under the unit's top plate. When a good-sized piece of magnetic material such as, for example, a cast-iron skillet is placed in the magnetic field of the induction coil is generating, the field transfers ("induces") energy into that metal. That transferred energy causes the metal the cooking vessel to become hot. By controlling the strength of the electromagnetic field, we can control the amount of heat being generated in the cooking vessel and we can change that amount instantaneously.

(To be technical, the field generates a loop current a flow of electricity within the metal of which the pot or pan is made, and that current flow through the resistance of the metal generates heat, just as current flowing through the resistance element of a conventional electric range's coil generates heat; the difference is that here, the heat is generated directly in the pot or pan itself, not in any part of the cooker.This type of current is called eddy current

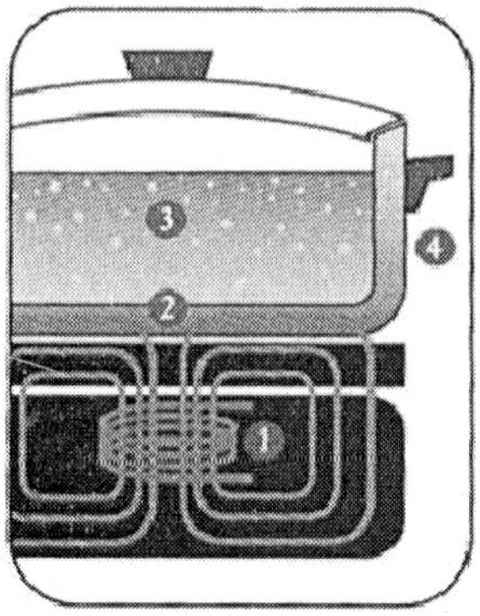

1. The element's electronics power a coil (the red lines) that produces a high-frequency electromagnetic field (represented by the orange lines).
2. That field penetrates the metal of the ferrous (magnetic-material) cooking vessel and sets up a circulating electric current, which generates heat. (But see the note below.)

3. The heat generated in the cooking vessel is transferred to the vessel's contents.
4. Nothing outside the vessel is affected by the field as soon as the vessel is removed from the element, or the element turned off, heat generation stops.

(Note: the process described at #2 above is called an "eddy current"; heat is also generated by another process called "hysteresis", which is the resistance of the ferrous material to rapid changes in magnetization. The relative contributions of the two effects is highly technical, with some sources emphasizing one and some the other but the general idea is unaffected: the heat is generated in the cookware.),

There is thus one point about induction: with current technology, induction cookers require that all your countertop cooking vessels be of a "ferrous" metal (one, such as iron,or steel that will readily sustain a magnetic field). Materials like aluminum, copper, and other metal are not usable on an induction cooker. But all that means is that you need iron or steel pots and pans. And that is no drawback in absolute terms, for it includes the best kinds of cookware in the world every top line is full of cookware of all sizes and shapes suitable for use on induction cookers (and virtually all of the lines will boast of it, because induction cooking is so popular with discerning cooks). Nor do you have to go to top-of-the-line names like All-Clad or Le Creuset, for many very reasonably priced cookware lines are also perfectly suited for induction cooking. But if you are considering induction cooking and have lot invested, literally or emotionally, in non-ferrous cookware, you do need to know the facts.

(And there are now available so-called "induction disks"-a round iron plate that will allow non-ferrous cookware to be used on an induction element; using such a disk loses many of the advantages of induction from high efficiency to no waste heat but those who want or need, say, a glass/pyrex or ceramic pot for some special use, it is possible to use it on an induction cooktop with such a disk.)But the life of induction cooker reduces drastically using a iron disk on the top plate.

MORE COOKING TECHNOLOGY IN FUTURE MAY COME BUT ITS TOO FAR.

On the horizon is newer technology that will apparently work with any metal cooking vessel, including copper and aluminum, but that technology though already being used in a few units of Japanese manufacturer is probably quite a few years away from maturity and from inclusion in most induction cookers. If you are interested in a new cooktop, it is, in our judgement, not worth waiting for that technology.

(The trick seems to be using a significantly high-frequency field, which is able to induce a current in any metal; ceramic and glass, however, would still be out of the running for cookware even when this new technology arrives if it ever does.)

Now finally on this side of that "horizon" is the so-called "zoneless" induction cooktop (every maker has its own trademarked term for "zoneless', but that's the common term). It rather seems, at this point, as if the mountain has labored to birth a mouse. The original promise of zoneless was a surface on which you could put down any size or shape of cooking vessel in any location or orientation and have everything work. In fact, even that, when once one comes to think it through, is not so very exciting except as it may involve cooking vessels of unusual sizes or shapes the few things (griddle, grill, fish pan) that don't function well, or at all, on standard circular heating elements; for most pots and pans, having fixed-location heating elements is just not an issue.

Well, zoneless has been in Europe for some years, and now is arrived in the U.S. with models from (as this is written) two BSH Group makers: Thermador and Gaggenau. (No doubt there will be more from other makers soon enough both the Electrolux Group and the Fagor Group have substantial expertise in "zoneless".) The issue on which, in our opinion, these new zoneless units disappoint is capacity: they are 36-inch units, a size one would normally expect to be able to carry up to five vessels but to achieve the "zoneless" quality, they restrict the cook to a maximum of four cooking vessels at any one time. To us, that seems a big step backward in technology.

It would be less disappointing were it not that there are now several units on the market that provide the option for true induction-powered "bridging" between a front-and-back element pair, effectively turning the two into a single quite long heating element, so that exactly the "problem" vessels grills, griddles, fish pans, and the like are accomodated perfectly well. There are 30-inch, four-element units with bridges, and 36-inch, five-element units as well. What one might gain over such a bridged unit by going to one that is 36 inches wide but only takes four vessels escapes us.

Finally, there is also now such a thing as an induction oven. (The usual heating coil on the base of the oven has been replaced by a ferrous plate, which is energized to heat by embedded induction coils beneath it so any sort of bakeware will work in it.) Expect to see more such things before long.

FACTS ABOUT LABOR INDUCTION

WHAT IS LABOR INDUCTION?

Sometimes it's safer to deliver your baby before you go into natural labor. Your doctor could use medicines to make your labor start "early," so you can deliver your baby vaginally. This is called "labor induction." Or it might be better for you not to go into labor at all and have a cesarean section (also called a "C" section).

WHY WOULD MY DOCTOR RECOMMEND LABOR INDUCTION?

The most common reason for labor induction is that the pregnancy has gone two weeks or more past the due date. The baby may get too big if you carry it this far past your due date. It may not be able to get enough food from inside your body. Your doctor might also recommend labor induction if:

- Your water breaks before you go into labor.
- You have high blood pressure.
- You have a serious infection.
- You have diabetes.

Your doctor will only recommend labor induction to protect you or your baby.

HOW WILL MY DOCTOR INDUCE MY LABOR?

There are several ways to induce labor. Toward the end of pregnancy, the cervix (the opening to the uterus, or womb) gets soft. It may even open up a little. Your doctor will check to see if your cervix is getting soft and opening up. If it isn't, your doctor may put a medicine in your vagina near the cervix. The medicine helps your cervix get soft and open up.

Your doctor may also "break your water" or use a finger to separate your cervix from the membranes (tissues) around your baby's head. This often makes labor start.

Your doctor will watch you closely. When you're ready, your doctor will start giving you a medicine called oxytocin. Oxytocin will start your contractions and help them to get strong and regular.

ARE THERE ANY RISKS TO LABOR INDUCTION?

One risk is that the medicine might not work. If your labor won't start with medicine, you might need to have a cesarean section. Or the medicine might make your contractions too strong. If this happens, your doctor might stop the medicine and wait for a while, or do a cesarean section. If you're being induced because of medical problems, there might be other risks.

WHAT CAN I EXPECT DURING LABOR INDUCTION?

Sometimes it takes two or three days to induce labor, but it usually takes less time. It takes more time if you're being induced really early or if it's your first baby. Don't eat very much before you come to the hospital. Remember that the medicines for labor induction can give you very strong contractions and might upset your stomach. Tell your doctor if you need help with the pain. In most cases, labor induction goes well and you can deliver your baby vaginally.

This handout is provided to you by your family doctor and the American Academy of Family Physicians. Other health-related information is available from the AAFP online at http://familydoctor.org.

This information provides a general overview and may not apply to everyone. Talk to your family doctor to find out if this information applies to you and to get more information on this subject.

LABOR INDUCTION

Labor induction is the process or treatment that stimulates childbirth and delivery. Inducing labor can be accomplished with pharmaceutical or non-pharmaceutical methods.

MEDICAL USES

Commonly accepted medical reasons for induction include:

- Postterm pregnancy, i.e. if the pregnancy has gone past the end of the 41st week.
- Intrauterine fetal growth restriction (IUGR).
- There are health risks to the woman in continuing the pregnancy (e.g. she has pre-eclampsia).
- Premature rupture of the membranes (PROM); this is when the membranes have ruptured, but labor does not start within a specific amount of time.
- Premature termination of the pregnancy (abortion).
- Fetal death in utero and previous history of stillbirth.
- Twin pregnancy continuing beyond 38 weeks.
- Previous health conditions that puts risk on the woman and/or her child such as diabetes, high blood pressure

Induction of labor in those who are either at or after term improves outcomes for the baby and decreases the number of C-sections performed.

METHODS OF INDUCTION

Methods of inducing labor include both pharmacological medication and mechanical or physical approaches.

Mechanical and physical approaches can include artificial rupture of membranes or membrane sweeping. The use of intrauterine catheters are also indicated. These work by compressing the cervix mechanically to generate release on prostaglandins in local tissues. There is no direct effect on the uterus.

Pharmacological methods are mainly using either dinoprostone (prostaglandin E2) or misoprostol (a prostaglandin E1 analogue)

MEDICATION

- Intravaginal, endocervical or extra-amniotic administration of prostaglandin, such as dinoprostone or misoprostol. Prostaglandin E2 is the most studied compound and with most evidence behind it. A range of different dosage forms are available with a variety of routes possible. The use of misoprostol has been extensively studied but normally in small, poorly defined studies. Only a very few countries have approved misoprostol for use in induction of labor.
- Intravenous administration of synthetic oxytocin preparations. A high dose does not seem to have greater benefits than a standard dose.
- Use of mifepristone has been described but is rarely used in practice.
- Relaxin has been investigated, but is not currently commonly used.
- mnemonic; ARNOP: Antiprogesterone, relaxin, nitric oxide donors, oxytocin, prostaglandins

NON-PHARMACEUTICAL

- "Membrane sweep", also known as membrane stripping, or "stretch and sweep" in Australia and the UK – during an internal examination, the practitioner moves their finger around the cervix to stimulate and/ or separate the membranes around the baby from the cervix. This causes a release of prostaglandins which can help to kick-start labor.
- Artificial rupture of the membranes (AROM or ARM) ("breaking the waters")
- Extra-amniotic saline infusion (EASI), in which a Foley catheter is inserted into the cervix and the distal portion expanded to dilate it and to release prostaglandins.
- Cook Medical Double Balloon known as the Cervical Ripening Balloon with Stylet for assisted placement is FDA approved. The Double balloon provides one balloon to be inflated with saline on one side of the Uterine side of the cervix and the second balloon to be inflated with saline on the vaginal side of the cervix.

WHEN TO INDUCE

The American Congress of Obstetricians and Gynecologists has recommended against elective induction before 41 weeks if there is no medical indication and the cervix is unfavorable. However, recent studies contradict this view. One recent study indicates that labor induction at term or post-term reduces the rate of caesarean section by 12%, and also reduces fetal death. On the other hand, observational/retrospective studies have shown that non-indicated, elective inductions before the 41st week of gestation are associated with an increased risk of requiring a caesarean section. Randomized clinical trials have not addressed this question. However, researchers have found that multiparous women who undergo labor induction without medical indicators are not predisposed to caesarean sections. Doctors and patients should have a discussion of risks and benefits when considering an induction of labor in the absence of an accepted medical indiction.

Studies have shown a slight increase in risk of infant mortality for births in the 41st and particularly 42nd week of gestation, as well as a higher risk of injury to the mother and child. Due to the increasing risks of advanced gestation, induction appears to reduce the risk for caesarean delivery after 41 weeks gestation and possibly earlier.

Inducing labor before 39 weeks in the absence of a medical indication (such as hypertension, IUGR, or pre-eclampsia) increases the risk of complications of prematurity including difficulties with respiration, infection, feeding, jaundice, neonatal intensive care unit admissions, and perinatal death.

Clinicians assess the odds of having a vaginal delivery after labor induction by a "Bishop Score". However, recent research has questioned the relationship between the Bishop score and a successful induction, finding that a poor Bishop score actually may improve the chance for a vaginal delivery after induction. A Bishop Score is done to assess the progression of the cervix prior to an induction. In order to do this, the cervix must be checked to see how much it has effaced, thinned out, and how far dilated it is. The score goes by a points system depending on five factors. Each factor is scored on a scale of either 0–2 or 0–3, any total score less than 5 holds a higher risk of delivering by caesarean section.

CRITICISMS OF INDUCTION

Induced labor may be more painful for the woman. This can lead to the increased use of analgesics and other pain-relieving pharmaceuticals. These interventions have been said to lead to an increased likelihood of caesarean section delivery for the baby. However, studies into this matter show differing results. One study indicated that while overall caesarean section rates from 1990–1997 remained at or below 20%, elective induction was associated with a doubling of the rate of caesarean section. Another study showed that elective induction in women who were not post-term increased a woman's chance of a C-section by two to three times. A more recent study indicated that induction may increase the risk of caesarean section if performed before the 40th week of gestation, but it has no effect or actually lowers the risk if performed after the 40th week.

The most recent reviews on the subject of induction and its effect on Cesaerean section indicate that there is no increase with induction and in fact there can be a reduction.

The Institute for Safe Medication Practices labeled pitocin a "high-alert medication" because of the high likelihood of "significant patient harm when it is used in error."Correspondingly, the improper use of pitocin is frequently an issue in malpractice litigation

Induction Training – How Does It Benefit the Company and the New Hire?

Various studies have shown that induction training directly affects employee attrition rate. It is a well-known fact that more than 25% of new employees decide to stay with or quit an organization in the first week of their stint. Employee training is very important for the organization, and delivering an effective induction training program to new employees who join the company should be its first priority.

The main purpose of induction training is to integrate new employees into the company and make them understand the systems and procedures followed by the organization. Induction training helps new employees settle down quickly in the new work environment, and gives them a sense of belonging.

Let's now look at the benefits of induction training, both from the perspective of the firm and the new employee.

BENEFITS OF EFFECTIVE INDUCTION TRAINING TO THE ORGANIZATION

1. Saves a lot of money and time: Induction training is the first training program in which the employee participates after he joins the organization. Induction training provides him all the information needed to start performing his duties. If an employee is trained well in the induction program, he can easily adapt to his new role and start delivering results quickly – that's how it saves the organization a lot of money and time.
2. Reduces employee turnover: People join a company with a lot of expectations, and at the same time, they have lot of questions about the organization. All these queries must be answered in the induction training. Ineffective induction training leaves new employee confused about the job. Employees may feel frustrated and helpless, if they are not trained properly. On the other hand, effective induction training goes a long way in increasing staff retention rate and reducing employee turnover in a big way.
3. Ensures operational efficiency: A good induction training program covers all aspects of the company thoroughly. It helps new employees become familiar with the organization's work culture, vision, mission, and goals. At the same time, new employees understand their own role in achieving the goals of the company. This will help enhance the efficiency of employees quickly, as they adjust to the work culture of the organization and get involved in their job. Overall, it greatly helps increase the operational efficiency of the organization.

ADVANTAGES OF PROPER INDUCTION TRAINING TO THE NEW EMPLOYEE

1. Makes the new employee feel respected and valued: Good induction training gives a warm welcome to the employee and focuses on clarifying all his doubts about the organization and his job. It ensures that the new employee feels comfortable in the organization. It also makes him feel that he is welcomed, respected, and valued. This motivates the new employee greatly.
2. Provides the necessary information: A comprehensive induction training program helps the new employee get all the necessary information about the company and clarifies the organization's expectations on him. This helps him understand the culture, work norms, policies and procedures of the organization, and thus enables him to quickly adapt to the work environment.
3. Helps in establishing good communication: Induction training helps the new employee in establishing good communication with the organization. As part of the induction training program, the new employee is introduced to his direct supervisor, other employees, leads, and directors of the organization. This makes him more comfortable when he has to communicate with them later.

Many employers see induction as a waste of valuable time but this is a critical process when you take on a new employee. Induction gives your new employee an objective view of your company, organisational culture, and work ethic, which will allow the employee to better integrate into the workplace.

WHAT SHOULD INDUCTION COVER?

The Advisory, Conciliation and Arbitration Service (ACAS) recommends that your induction programme should include:

- An introduction to your company/department and its personnel structure;
- A layout of the establishment, depending on your building size and structure;
- The terms and conditions of employment;
- Relevant personnel policies, such as training, promotion, and health and safety;
- Your company rules and procedures;
- Arrangements for your new employee's involvement in their business area or unit;
- Employee benefits or facilities; and
- An induction checklist, which will be used as a quick reference guide

Make sure you cover these essentials before you start the induction process

- As the manager, supervisor or HR representative, you must be part of the induction process;

- Ensure that your new employee's office and office equipment is setup for them when they arrive. It will create a good impression and avoid the wasted hours;
- Notify all your colleagues, at least a week before your new employee arrives, that a new member will be joining the team; and
- Contact your new employee to arrange:
- When he must arrive at the office;
- Where he must park;
- How he will access the building;
- What your company's dress code is; and
- What he must bring to work on the first day.

LET'S LEARN FROM UKZN AND UNISA'S INDUCTION PROCESSES

Local universities, such as the University of Kwa-Zulu Natal (UKZN) and the University of South Africa (Unisa), enforce strict procedures within the recruitment and induction processes because a good induction programme will benefit both the university and their new employee.

WHAT ARE THE BENEFITS OF AN INDUCTION PROGRAMME?

UKZN and Unisa both agree that the benefits of induction programmes are that these:

- Create a positive perception of the university while communicating its established culture, values and goals;
- Increase the retention of new employees as they are more likely to commit to a longer term at the university;
- Ensure operational efficiency as new employees become productive sooner; and
- Ensuring a happy and motivated workforce.

4 benefits that your new employee will get from an induction programme

1. The employee will feel welcomed into your organisation, be respected and be made to feel more comfortable in the workplace. He will also find it easier to integrate into the workplace;
2. Your new employee will feel that he made the right decision to join your organisation;
3. A good induction programme will help to build the new employee's self-esteem, morale and sense of motivation; and
4. An induction programme will establish good communication between you and/or the supervisor and your new employee from the very beginning.

3

Maxwell's Law

EARLY LIFE, 1831–1839

James Clerk Maxwell was born on 13 June 1831 at 14 India Street, Edinburgh, to John Clerk Maxwell of Middlebie, an advocate, and Frances Cay daughter of Robert Hodshon Cay and sister of John Cay. (His birthplace now houses a museum operated by the James Clerk Maxwell Foundation.) His father was a man of comfortable means of the Clerk family of Penicuik, holders of the baronetcy of Clerk of Penicuik. His father's brother was the 6th Baronet. He had been born "John Clerk", adding the surnameMaxwell to his own after he inherited (as an infant in 1793) the Middlebie country estate near Corsock, Kirkcudbrightshire, from connections to the Maxwell family, themselves members of the peerage. James was a first cousin of both the artist Jemima Blackburn (the daughter of his father's sister) and the civil engineer William Dyce Cay (the son of his mother's brother). Cay and Maxwell were close friends and Cay acted as his best man when Maxwell married.

Maxwell's parents met and married when they were well into their thirties; his mother was nearly 40 when he was born. They had had one earlier child, a daughter named Elizabeth, who died in infancy.

When Maxwell was young his family moved to Glenlair House, which his parents had built on the 1,500 acres (610 ha) Middlebie estate. All indications suggest that Maxwell had maintained an unquenchable curiosity from an early age. By the age of three, everything that moved, shone, or made a noise drew the question: "what's the go o' that?" In a passage added to a letter from his father to his sister-in-law Jane Cay in 1834, his mother described this innate sense of inquisitiveness:

He is a very happy man, and has improved much since the weather got moderate; he has great work with doors, locks, keys, etc., and "show me how it doos" is never out of his mouth. He also investigates the hidden course of streams and bell-wires, the way the water gets from the pond through the wall....

EDUCATION, 1839–1847

Recognising the potential of the young boy, Maxwell's mother Frances took responsibility for James's early education, which in the Victorian era was largely the job of the woman of the house. At eight he could recite long passages of Milton and the whole of the 119th psalm (176 verses). Indeed, his knowledge of scripture was already detailed; he could give chapter and verse for almost any quotation from the psalms. His mother was taken ill with abdominal cancer and, after an unsuccessful operation, died in December 1839 when he was eight years old. His education was then overseen by his father and his father's sister-in-law Jane, both of whom played pivotal roles in his life. His formal schooling began unsuccessfully under the guidance of a 16 year old hired tutor. Little is known about the young man hired to instruct Maxwell, except that he treated the younger boy harshly, chiding him for being slow and wayward. The tutor was dismissed in November 1841 and, after considerable thought, Maxwell was sent to the prestigious Edinburgh Academy. He lodged during term times at the house of his aunt Isabella. During this time his passion for drawing was encouraged by his older cousin Jemima.

Edinburgh Academy, where Maxwell was schooled.

The 10 year old Maxwell, having been raised in isolation on his father's countryside estate, did not fit in well at school. The first year had been full, obliging him to join the second year with classmates a year his senior. His mannerisms and Galloway accent struck the other boys as rustic. Having arrived on his first day of school wearing a pair of homemade shoes and a tunic, he earned the unkind nickname of "Daftie". He never seemed to resent the epithet, bearing it without complaint for many years. Social isolation at the Academy ended when he met Lewis Campbell and Peter Guthrie Tait, two boys of a similar age who were to become notable scholars later in life. They remained lifelong friends.

Maxwell was fascinated by geometry at an early age, rediscovering the regular polyhedra before he received any formal instruction.Despite winning the school's scripture biography prize in his second year, his academic work remained unnoticed until, at the age of 13, he won the school's mathematical medal and first prize for both English and poetry.

Maxwell's interests ranged far beyond the school syllabus and he did not pay particular attention to examination performance. He wrote his first scientific paper at the age of 14. In it he described a mechanical means of drawing mathematical curves with a piece of twine, and the properties of ellipses, Cartesian ovals, and related curves with more than two foci. His work "Oval Curves" was presented to the Royal Society of Edinburgh by James Forbes, a professor of natural philosophy at the University of Edinburgh, because Maxwell was deemed too young to present the work himself. The work was not entirely original, since René Descartes had also examined the properties of such multifocal ellipses in the 17th century, but he had simplified their construction.

UNIVERSITY OF EDINBURGH, 1847–1850

Old College, University of Edinburgh

Maxwell left the Academy in 1847 at age 16 and began attending classes at the University of Edinburgh. He had the opportunity to attend the University of Cambridge, but decided, after his first term, to complete the full course of his undergraduate studies at Edinburgh. The academic staff of the University included some highly regarded names; his first year tutors included Sir William Hamilton, who lectured him on logic and metaphysics, Philip Kelland on mathematics, and James Forbes on natural philosophy. He did not find his classes at the University demanding, and was therefore able to immerse himself in private study during free time at the University and particularly when back home at Glenlair. There he would experiment with improvised chemical, electric, and magnetic apparatus, however his chief concerns regarded the properties of polarised light. He constructed shaped blocks of gelatine, subjected them to various stresses, and with a pair of polarising prisms given to him by William Nicol, viewed the coloured fringes that had

paper "Experiments on Colour" laid out the principles of colour combination and was presented to the Royal Society of Edinburgh in March 1855. Maxwell was this time able to deliver it himself. Maxwell was made a fellow of Trinity on 10 October 1855, sooner than was the norm, and was asked to prepare lectures on hydrostatics and optics and to set examination papers. The following February he was urged by Forbes to apply for the newly vacant Chair of Natural Philosophy at Marischal College, Aberdeen. His father assisted him in the task of preparing the necessary references, but died on 2 April at Glenlair before either knew the result of Maxwell's candidacy. Maxwell accepted the professorship at Aberdeen, leaving Cambridge in November 1856.

MARISCHAL COLLEGE, ABERDEEN, 1856–1860

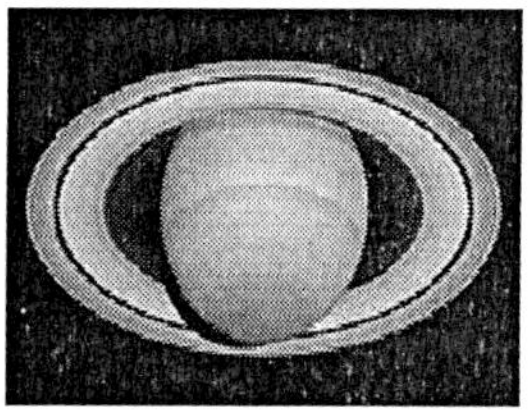

Maxwell proved that the Rings of Saturn were made of numerous small particles.

The 25-year-old Maxwell was a good 15 years younger than any other professor at Marischal. He engaged himself with his new responsibilities as head of a department, devising the syllabus and preparing lectures. He committed himself to lecturing 15 hours a week, including a weekly pro bono lecture to the local working men's college. He lived in Aberdeen during the six months of the academic year and spent the summers at Glenlair, which he had inherited from his father.

James and Katherine Maxwell, 1869

He focused his attention on a problem that had eluded scientists for 200 years: the nature of Saturn's rings. It was unknown how they could remain stable without breaking up, drifting away or crashing into Saturn.The problem took on a particular resonance at that time because St John's College, Cambridge had chosen it as the topic for the 1857 Adams Prize. Maxwell devoted two years to studying the problem, proving that a regular solid ring could not be stable, while a fluid ring would be forced by wave action to break up into blobs. Since neither was observed, Maxwell concluded that the rings must be composed of numerous small particles he called "brick-bats", each independently orbiting Saturn. Maxwell was awarded the £130 Adams Prize in 1859 for his essay "On the stability of the motion of Saturn's rings"; he was the only entrant to have made enough headway to submit an entry. His work was so detailed and convincing that when George Biddell Airy read it he commented "It is one of the most remarkable applications of mathematics to physics that I have ever seen." It was considered the final word on the issue until direct observations by the Voyager flybys of the 1980s confirmed Maxwell's prediction.

In 1857 Maxwell befriended the Reverend Daniel Dewar, who was then the Principal of Marischal. Through him Maxwell met Dewar's daughter, Katherine Mary Dewar. They were engaged in February 1858 and married in Aberdeen on 2 June 1858. On the marriage record, Maxwell is listed as Professor of Natural Philosophy in Marischal College, Aberdeen. Seven years Maxwell's senior, comparatively little is known of Katherine, although it is known that she helped in his lab and worked on experiments in viscosity. Maxwell's biographer and friend, Lewis Campbell, adopted an uncharacteristic reticence on the subject of Katherine, though describing their married life as "one of unexampled devotion". In 1860 Marischal College merged with the neighbouring King's College to form the University of Aberdeen. There was no room for two professors of Natural Philosophy, so Maxwell, despite his scientific reputation, found himself laid off. He was unsuccessful in applying for Forbes's recently vacated chair at Edinburgh, the post instead going to Tait. Maxwell was granted the Chair of Natural Philosophy at King's College, London, instead. After recovering from a near-fatal bout of smallpox in 1860, Maxwell moved to London with his wife.

KING'S COLLEGE, LONDON, 1860–1865

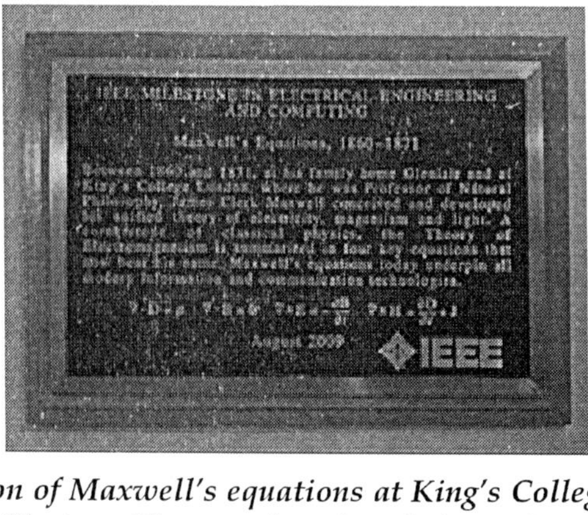

Commemoration of Maxwell's equations at King's College. One of three identical IEEE Milestone Plaques, the others being at Maxwell's birthplace in Edinburgh and the family home at Glenlair.

Maxwell's time at King's was probably the most productive of his career. He was awarded the Royal Society'sRumford Medal in 1860 for his work on colour and was later elected to the Society in 1861. This period of his life would see him display the world's first light-fast colour photograph, further develop his ideas on the viscosityof gases, and propose a system of defining physical quantities—now known as dimensional analysis. Maxwell would often attend lectures at the Royal Institution, where he came into regular contact with Michael Faraday. The relationship between the two men could not be described as being close, because Faraday was 40 years Maxwell's senior and showed signs of senility. They nevertheless maintained a strong respect for each other's talents.

Blue plaque, 16 Palace Gardens Terrace, Kensington, Maxwell's home, 1860–1865

This time is especially noteworthy for the advances Maxwell made in the fields of electricity and magnetism. He examined the nature of both electric and magnetic fields in his two-part paper "On physical lines of force", which was published in 1861. In it he provided a conceptual model for electromagnetic induction, consisting of tiny spinning cells of magnetic flux. Two more parts were later added to and published in that same paper in early 1862. In the first additional part he discussed the nature of electrostatics and displacement current. In the second additional part, he dealt with the rotation of the plane of the polarisation of light in a magnetic field, a phenomenon that had been discovered by Faraday and is now known as the Faraday effect.

LATER YEARS, 1865–1879

The gravestone at Parton Kirk (Galloway) of James Clerk Maxwell, his parents and his wife

This memorial stone to James Clerk Maxwell stands on a green in front of the church, beside the war memorial at Parton (Galloway).

In 1865 Maxwell resigned the chair at King's College, London, and returned to Glenlair with Katherine. In his paper "On reciprocal figures, frames and diagrams of forces" (1870) he discussed the rigidity of various designs of lattice. He wrote the textbook Theory of Heat(1871) and the treatise Matter and Motion (1876). Maxwell was also the first to make explicit use of dimensional analysis, in 1871.

In 1871 he returned to Cambridge to become the first Cavendish Professor of Physics. Maxwell was put in charge of the development of the Cavendish Laboratory, supervising every step in the progress of the building and of the purchase of the collection of apparatus. One of Maxwell's last great contributions to science was the editing (with copious original notes) of the research of Henry Cavendish, from which it appeared that Cavendish researched, amongst other things, such questions as the density of the Earth and the composition of water.

Maxwell died in Cambridge of abdominal cancer on 5 November 1879 at the age of 48. His mother had died at the same age of the same type of cancer. The minister who regularly visited him in his last weeks was astonished at his lucidity and the immense power and scope of his memory, but comments more particularly,

... his illness drew out the whole heart and soul and spirit of the man: his firm and undoubting faith in the Incarnation and all its results; in the full sufficiency of the Atonement; in the work of the Holy Spirit. He had gauged and fathomed all the schemes and systems of philosophy, and had found them utterly empty and unsatisfying — "unworkable" was his own word about them — and he turned with simple faith to the Gospel of the Saviour.

As death approached Maxwell told a Cambridge colleague

I have been thinking how very gently I have always been dealt with. I have never had a violent shove all my life. The only desire which I can have is like David to serve my own generation by the will of God, and then fall asleep.

Maxwell is buried at Parton Kirk, near Castle Douglas in Galloway close to where he grew up. The extended biography The Life of James Clerk Maxwell, by his former schoolfellow and lifelong friend Professor Lewis Campbell, was published in 1882. His collected works were issued in two volumes by the Cambridge University Press in 1890.

PERSONAL LIFE

As a great lover of Scottish poetry, Maxwell memorised poems and wrote his own. The best known is Rigid Body Sings, closely based on "Comin' Through the Rye" by Robert Burns, which he apparently used to sing while accompanying himself on a guitar. It has the opening lines

GIN A BODY MEET A BODY

Flyin' through the air. Gin a body hit a body, Will it fly? And where?

A collection of his poems was published by his friend Lewis Campbell in 1882.

Descriptions of Maxwell remark upon his remarkable intellectual qualities being matched by social awkwardness.

Maxwell was an evangelical Presbyterian and in his later years became an Elder of the Church of Scotland. Maxwell's religious beliefs and related activities have been the focus of a number of papers. Attending both Church of Scotland (his father's denomination) and Episcopalian (his mother's denomination) services as a child, Maxwell later underwent an evangelical conversion in April 1853. One facet of this conversion may have aligned him with an antipositivist position.

SCIENTIFIC LEGACY

ELECTROMAGNETISM

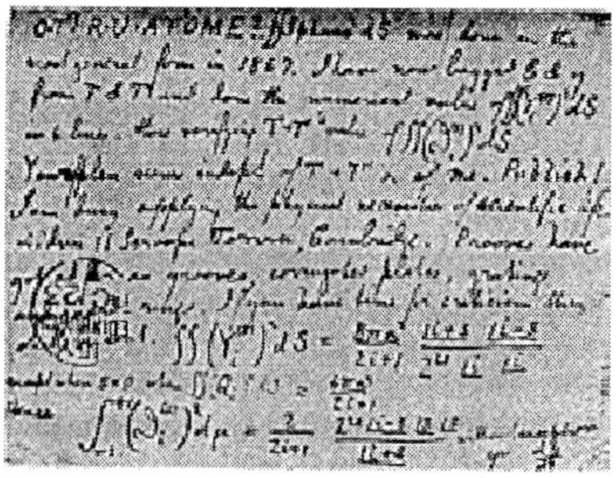

A postcard from Maxwell to Peter Tait

Maxwell had studied and commented on electricity and magnetism as early as 1855 when his paper "On Faraday's lines of force" was read to the Cambridge Philosophical Society. The paper presented a simplified model of Faraday's work and how electricity and magnetism are related. He reduced all of the current knowledge into a linked set of differential equations with 20 equations in 20 variables. This work was later published as "On Physical Lines of Force" in March 1861.

Around 1862, while lecturing at King's College, Maxwell calculated that the speed of propagation of an electromagnetic field is approximately that of the speed of light (see speed of light#electromagnetic constants). He considered this to be more than just a coincidence, commenting, "We can scarcely avoid the conclusion that light consists in the transverse undulations of the same medium which is the cause of electric and magnetic phenomena."

Working on the problem further, Maxwell showed that the equations predict the existence of waves of oscillating electric and magnetic fieldsthat travel through empty space at a speed that could be predicted from simple electrical experiments; using the data available at the time, Maxwell obtained a velocity of 310,740,000 metres per second (1.0195×10 ft/s). In his 1864 paper "A Dynamical Theory of the Electromagnetic Field", Maxwell wrote, "The agreement of the results seems to show that light and magnetism are affections of the same substance, and that light is an electromagnetic disturbance propagated through the field according to electromagnetic laws".

His famous twenty equations, in their modern form of four partial differential equations, first appeared in fully developed form in his textbook A Treatise on Electricity and Magnetismin 1873. Most of this work was done by Maxwell at Glenlair during the period between holding his London post and his taking up the Cavendish chair. Maxwell expressed electromagnetism in the algebra of quaternions and made the electromagnetic potential the centrepiece of his theory. In 1881 Oliver Heaviside replaced Maxwell's electromagnetic potential field by 'force fields' as the centrepiece of electromagnetic theory. Heaviside reduced the complexity of Maxwell's theory down to four differential equations, known now collectively as Maxwell's Laws or Maxwell's equations. According to Heaviside, the electromagnetic potential field was arbitrary and needed to be "murdered". The use of scalar and vector potentials is now standard in the solution of Maxwell's equations.

A few years later there was a debate between Heaviside and Peter Guthrie Tait about the relative merits of vector analysis and quaternions. The result was the realisation that there was no need for the greater physical insights provided by quaternions if the theory was purely local, and vector analysis became commonplace. Maxwell was proven correct, and his quantitative connection between light and electromagnetism is considered one of the great accomplishments of 19th century mathematical physics.

Maxwell also introduced the concept of the electromagnetic field in comparison to force lines that Faraday described. By understanding the propagation of electromagnetism as a field emitted by active particles, Maxwell could advance his work on light. At that time, Maxwell believed that the propagation of light required a medium for the waves, dubbed the luminiferous aether. Over time, the existence of such a medium, permeating all space and yet apparently undetectable by mechanical means, proved impossible to reconcile with experiments such as the Michelson–Morley experiment. Moreover, it seemed to require an absolute frame of reference in which the equations were valid, with the distasteful result that the equations changed form for a moving observer. These difficulties inspired Albert Einstein to formulate the theory of special relativity; in the process Einstein dispensed with the requirement of a stationary luminiferous aether.

In electromagnetism, one of the fundamental fields of physics, the introduction of Maxwell's equations (mainly in "A Dynamical Theory of the Electromagnetic Field") was one of the most important aggregations of empirical facts in the history of physics. It took place in the nineteenth century, starting from basic experimental observations, and leading to the formulations of numerous mathematical equations, notably by Charles-Augustin de Coulomb, Hans Christian Ørsted, Carl Friedrich Gauss, Jean-Baptiste Biot, Félix Savart, André-Marie Ampère, and Michael Faraday. The apparently disparate laws and phenomena of electricity and magnetism were integrated by James Clerk Maxwell, who published an early form of the equations, which modify Ampère's circuital law by introducing a displacement current term. He showed that these equations imply that light propagates as electromagnetic waves. His laws were reformulated by Oliver Heaviside in the more modern and compact vector calculus formalism he independently developed. Increasingly powerful mathematical descriptions of the electromagnetic field were developed, continuing into the twentieth century, enabling the equations to take on simpler forms by advancing more sophisticated mathematics.

RELATIONSHIPS AMONG ELECTRICITY, MAGNETISM, AND THE SPEED OF LIGHT

The relationships among electricity, magnetism, and the speed of light can be summarized by the modern equation:{\displaystyle c={\frac {1}{\sqrt {\mu _{0}\varepsilon _{0}}}}\. }

The left-hand side is the speed of light and the right-hand side is a quantity related to the constants that appear in the equations governing electricity and magnetism. Although the right-hand side has units of velocity, it can be inferred from measurements of electric and magnetic forces, which involve no physical velocities. Therefore, establishing this relationship provided convincing evidence that light is an electromagnetic phenomenon.

The discovery of this relationship started in 1855, when Wilhelm Eduard Weber and Rudolf Kohlrausch determined that there was a quantity related to electricity and magnetism, "the ratio of the absolute electromagnetic unit of charge to the absolute electrostatic unit of charge" (in modern language, the value))))))))and determined that it should have units of velocity. They then measured this ratio by an experiment which involved charging and discharging a Leyden jar and measuring the magnetic force from the discharge current, and found a value 3.107×10 m/s, remarkably close to the speed of light, which had recently been measured at 3.14×10 m/s by Hippolyte Fizeau in 1848 and at 2.98×10 m/s by Léon Foucault in 1850. However, Weber and Kohlrausch did not make the connection to the speed of light. Towards the end of 1861 while working on part III of his paper On Physical Lines of Force, Maxwell travelled from Scotland to London and looked up Weber and Kohlrausch's results. He converted them into a format which was compatible with his own writings, and in doing so he established the connection to the speed of light and concluded that light is a form of electromagnetic radiation.

Maxwell's equations are a set of partial differential equations that, together with the Lorentz force law, form the foundation of classical electromagnetism, classical optics, and electric circuits. They underpin all electric, optical and radio technologies, including power generation, electric motors, wireless communication, cameras, televisions, computers etc. Maxwell's equations describe how electric and magnetic fields are generated by charges, currents, and changes of each other. One important consequence of the equations is that they demonstrate how fluctuating electric and magnetic fields propagate at the speed of light. Known as electromagnetic radiation, these waves may occur at various wavelengths to produce a spectrum from radio waves to ã-rays. The equations are named after the physicist and mathematician James Clerk Maxwell, who between 1861 and 1862 published an early form of the equations, and first proposed that light is an electromagnetic phenomenon.

The equations have two major variants. The microscopic Maxwell equations have universal applicability, but are unwieldy for common calculations. They relate the electric and magnetic fields to total charge and total current, including the complicated charges and currents in materials at the atomic scale. The "macroscopic" Maxwell equations define two new auxiliary fields that describe the large-scale behaviour of matter without having to consider atomic scale details. However, their use requires experimentally determining parameters for a phenomenological description of the electromagnetic response of materials.

THE TERM MAXWELL'S EQUATIONS

The four modern Maxwell's equations can be found individually throughout his 1861 paper, derived theoretically using a molecular vortex model of Michael Faraday's "lines of force" and in conjunction with the experimental result of Weber and Kohlrausch. But it wasn't until 1884 that Oliver Heaviside, concurrently with similar work by Josiah Willard Gibbs and Heinrich Hertz, grouped the twenty equations together into a set of only four, via vector notation. This group of four equations was known variously as the Hertz–Heaviside equations and the Maxwell–Hertz equations, but are now universally known as Maxwell's equations. Heaviside's equations, which are taught in textbooks and universities as Maxwell's equations are not exactly the same as the ones due to Maxwell, and, in fact, the latter are more easily forced into the mold of quantum physics. This very subtle and paradoxical sounding situation can perhaps be most easily understood in terms of the similar situation that exists with respect to Newton's second law of motion. In textbooks and in classrooms the law F = ma is attributed to Newton, but his second law was in fact F = p′, where p′ is the time derivative of the momentum p. This seems a trivial enough fact until you realize that F = p′ remains true in the context of Special relativity. The equation F = p′ is clearly visible in a glass case in the Wren Library of Trinity College, Cambridge, where Newton's manuscript is open to the relevant page.

The term "Maxwell's equations" is often used for equivalent alternative formulations. Versions of Maxwell's equations based on the electric and magnetic potentials are preferred for explicitly solving the equations as a boundary value problem, analytical mechanics, or for use in quantum mechanics. The spacetime formulations (i.e., on spacetime rather than space and time separately), are commonly used in high energy and gravitational physics because they make the compatibility of the equations with special and general relativity manifest. In fact, Einstein developed special and general relativity to accommodate the absolute speed of light that drops out of the Maxwell equations with the principle that only relative movement has physical consequences.

Since the mid-20th century, it has been understood that Maxwell's equations are not exact, but a classical field theoryapproximation of some aspects of the fundamental theory of quantum electrodynamics, although some quantum features, such as quantum entanglement, are completely absent and in no way approximated. (For example, quantum cryptography has no approximate version in Maxwell theory.) In many situations, though, deviations from Maxwell's equations are immeasurably small. Exceptions include nonclassical light, photon–photon scattering, quantum optics, and many other phenomena related to photons or virtual photons.

Maxwell's contribution to science in producing these equations lies in the correction he made to Ampère's circuital law in his 1861 paper On Physical Lines of Force. He added the displacement current term to Ampère's circuital law and this enabled him to derive the electromagnetic wave equation in his later 1865 paper A Dynamical Theory of the Electromagnetic Field and to demonstrate the fact that light is an electromagnetic wave. This fact was later confirmed experimentally by Heinrich Hertz in 1887. The physicist Richard Feynman predicted that, "From a long view of the history of mankind, seen from, say, ten thousand years from now, there can be little doubt that the most significant event of the 19th century will be judged as Maxwell's discovery of the laws of electrodynamics. The American Civil War will pale into provincial insignificance in comparison with this important scientific event of the same decade."

The concept of fields was introduced by, among others, Faraday. Albert Einstein wrote:

The precise formulation of the time-space laws was the work of Maxwell. Imagine his feelings when the differential equations he had formulated proved to him that electromagnetic fields spread in the form of polarised waves, and at the speed of light! Too few men in the world has such an experience been vouchsafed... it took physicists some decades to grasp the full significance of Maxwell's discovery, so bold was the leap that his genius forced upon the conceptions of his fellow workers.

Heaviside worked to eliminate the potentials (electric potential and magnetic potential) that Maxwell had used as the central concepts in his equations; this effort was somewhat controversial, though it was understood by 1884 that the potentials must propagate at the speed of light like the fields, unlike the concept of instantaneous action-at-a-distance like the then conception of gravitational potential.

ON PHYSICAL LINES OF FORCE

The four equations we use today appeared separately in Maxwell's 1861 paper, On Physical Lines of Force:

1. Equation (56) in Maxwell's 1861 paper is " • B = 0.
2. Equation (112) is Ampère's circuital law, with Maxwell's addition of displacement current. This may be the most remarkable contribution of Maxwell's work, enabling him to derive the electromagnetic wave equation in his 1865 paper A Dynamical Theory of the Electromagnetic Field, showing that light is an electromagnetic wave. This lent the equations their full significance with respect to understanding the nature of the phenomena he elucidated. (Kirchhoff derived the telegrapher's equations in 1857 without using displacement current, but he did use Poisson's equation and the equation of continuity, which are the mathematical ingredients of the displacement current. Nevertheless, believing his equations to be applicable only inside an electric wire, he cannot be credited with the discovery that light is an electromagnetic wave).
3. Equation (115) is Gauss's law.
4. Equation (54) expresses what Oliver Heaviside referred to as 'Faraday's law', which addresses the time-variant aspect of electromagnetic induction, but not the one induced by motion; Faraday's original flux law accounted for both. Maxwell deals with the motion-related aspect of electromagnetic induction, $v \times B$, in equation (77), which is the same as equation (D) in Maxwell's original equations as listed below. It is expressed today as the force law equation, $F = q(E + v \times B)$, which sits adjacent to Maxwell's equations and bears the name Lorentz force, even though Maxwell derived it when Lorentz was still a young boy.

The difference between the B and the H vectors can be traced back to Maxwell's 1855 paper entitled On Faraday's Lines of Force which was read to the Cambridge Philosophical Society. The paper presented a simplified model of Faraday's work, and how the two phenomena were related. He reduced all of the current knowledge into a linked set of differential equations.

Figure of Maxwell's molecular vortex model. For a uniform magnetic field, the field lines point outward from the display screen, as can be observed from the black dots in the middle of the hexagons. The vortex of each hexagonal molecule rotates counter-clockwise. The small green circles are clockwise rotating particles sandwiching between the molecular vortices.

It is later clarified in his concept of a sea of molecular vortices that appears in his 1861 paper On Physical Lines of Force. Within that context, H represented pure vorticity (spin), whereas B was a weighted vorticity that was weighted for the density of the vortex sea. Maxwell considered magnetic permeability μ to be a measure of the density of the vortex sea. Hence the relationship,

1. Magnetic induction current causes a magnetic current density B = ì H was essentially a rotational analogy to the linear electric current relationship,
2. Electric convection current J = ñ v where ñ is electric charge density. B was seen as a kind of magnetic current of vortices aligned in their axial planes, with H being the circumferential velocity of the vortices. With μ representing vortex density, it follows that the product of μ with vorticity H leads to the magnetic field denoted as B.

The electric current equation can be viewed as a convective current of electric charge that involves linear motion. By analogy, the magnetic equation is an inductive current involving spin. There is no linear motion in the inductive current along the direction of the B vector. The magnetic inductive current represents lines of force. In particular, it represents lines of inverse square lawforce.

The extension of the above considera.ions confirms that where B is to H, and where J is to ñ, then it necessarily follows from Gauss's law and from the equation of continuity of charge that E is to D. i.e. B parallels with E, whereas H parallels with D.

A DYNAMICAL THEORY OF THE ELECTROMAGNETIC FIELD

In 1865 Maxwell published A Dynamical Theory of the Electromagnetic Field in which he showed that light was an electromagnetic phenomenon. Confusion over the term "Maxwell's equations" sometimes arises because it has been used for a set of eight equations that appeared in Part III of Maxwell's 1865 paper A Dynamical Theory of the Electromagnetic Field, entitled "General Equations of the Electromagnetic Field", and this confusion is compounded by the writing of six of those eight equations as three separate equations (one for each of the Cartesian axes), resulting in twenty equations and twenty unknowns. (As noted above, this terminology is not common: Modern use of the term "Maxwell's equations" refer to the Heaviside restatements.)

A TREATISE ON ELECTRICITY AND MAGNETISM

In A Treatise on Electricity and Magnetism, an1873 treatise on electromagnetism written by James Clerk Maxwell, eleven general equations of the electromagnetic field are listed and these include the eight that are listed in the 1865 paper.

RELATIVITY

HISTORY OF SPECIAL RELATIVITY

Maxwell's equations were an essential inspiration for Einstein's development of special relativity. Possibly the most important aspect was their denial of instantaneous action at a distance. Rather, according to them, forces are propagated at the velocity of light through the electromagnetic field.

Maxwell's original equations are based on the idea that light travels through a sea of molecular vortices known as the "luminiferous aether", and that the speed of light has to be respective to the reference frame of this aether. Measurements designed to measure the speed of the Earth through the aether conflicted with this notion, though.

A more theoretical approach was suggested by Hendrik Lorentz along with George FitzGerald and Joseph Larmor. Both Larmor (1897) and Lorentz (1899, 1904) derived the Lorentz transformation (so named by Henri Poincaré) as one under which Maxwell's equations were invariant. Poincaré (1900) analyzed the coordination of moving clocks by exchanging light signals. He also established the mathematical group property of the Lorentz transformation (Poincaré 1905). Sometimes this transformation is called the FitzGerald–Lorentz transformation or even the FitzGerald–Lorentz–Einstein transformation.

Albert Einstein dismissed the notion of the aether as an unnecessary one, and he concluded that Maxwell's equations predicted the existence of a fixed speed of light, independent of the velocity of the observer. Hence, he used the Maxwell's equations as the starting point for his Special Theory of Relativity. In doing so, he established that the FitzGerald–Lorentz transformation is valid for all matter and space, and not just Maxwell's equations. Maxwell's equations played a key role in Einstein's groundbreaking scientific paper on special relativity (1905). For example, in the opening paragraph of his paper, he began his theory by noting that a description of an electric conductor moving with respect to a magnet must generate a consistent set of fields regardless of whether the force is calculated in the rest frame of the magnet or that of the conductor.

The general theory of relativity has also had a close relationship with Maxwell's equations. For example, Theodor Kaluza and Oskar Klein in the 1920s showed that Maxwell's equations could be derived by extending general relativity into five physical dimensions. This strategy of using additional dimensions to unify different forces remains an active area of research in physics.

This paper is written in commemoration of the 150th anniversary of James Clerk Maxwell's taking up, in 1856, the chair of natural philosophy in Marischal College, Aberdeen (later, in 1860, to merge with King's College to form the University of Aberdeen). It is the text of a lecture given by the author at a meeting held in Maxwell's honour in Aberdeen in September 2006. It should be added that this latter year is also the 175th anniversary of his birth, on 13 June 1831, in Edinburgh.

Maxwell's theory of the electromagnetic field is one of the main pillars of modern theoretical physics and, of course, played a key role in the formulation and the development of Einstein's special theory of relativity. The purpose of this article is to give a brief description of the events that led up to Maxwell's (1865) paper and his 1873 Treatise and which gave to the world the conclusions of one of the most fundamental pieces of scientific research ever undertaken. It will take the form of a brief history of the developments of electricity and magnetism from ancient times until 1905 when Einstein published his special theory of relativity. In keeping with the intention that the lecture and subsequent paper should be for a lay scientific audience, the mathematical details will be kept to a minimum. However, some mathematical language will be needed to aid coherency, especially in the later stages, but it is hoped that the paper can be read, with some profit, by those who wish or have to disregard such details.

The theory of the electromagnetic field, especially in its more advanced stages, has a complex history and many interesting sideshows must necessarily be omitted. However, the main developments will be included and, it is hoped, set in the correct context. The history of this subject, and thus the content of this paper, will be divided into five parts: first, the knowledge of the ancients, culminating in the work of Gilbert; second, the experimental period from Gilbert to the beginning of the nineteenth century; third, the short but exciting experimental and theoretical period from Oersted to the conclusion of the researches of Faraday; fourth, the period of Maxwell; and fifth, the period leading from Maxwell's great treatise to Einstein's special theory of relativity.

EARLY DAYS

The original source of awareness in matters, electrical or magnetic, no doubt comes from natural causes such as the discharge of lightning or the pale glow sometimes seen on the sharp ends of certain objects such as ships' masts or even warriors' spears, during stormy weather. However, as a topic of scientific debate, it appears to have come from two naturally occurring materials, that is, amber and loadstone. Amber, a resin from a type of pine tree, often found in the appearance of stone washed up in certain coastal locations, was known to have the special property that, when rubbed, it had the power to attract a variety of light objects, such as pieces of straw or fur placed in its vicinity. Amber and its properties were known to the ancient Greek and Roman civilizations. Loadstone (or lodestone) is a naturally occurring magnetized piece of the iron ore called magnetite, whose ability to attract iron was also well known to these cultures. Thales of Miletus apparently experimented with amber and loadstone in ca 600 BC. The ancient Chinese civilization may well have known of loadstone much earlier than the Europeans. It has also been known, probably for at least 1000 years, that a long thin piece of loadstone will, if rested on a piece of wood which is then floated on water, turn so that the tips of its long axis point north and south. Thus, the mariner's compass was born and was known to the Chinese, Muslim and north European sailors. Although the respective properties of amber and loadstone contained many similarities, their basic difference, that amber attracted many different types of object whereas loadstone attracted only iron, seemed sufficient to keep them separated.

There is little of significance to report further on the knowledge of amber and loadstone until the scientific renaissance came to Europe. Then in the middle of the sixteenth century, the Italian, Jerome Cardan, promoted further the case for the difference in amber and loadstone by carefully listing and contrasting their respective properties. He stressed, as mentioned above, that amber would attract anything sufficiently light whereas loadstone attracted only iron. He also made the interesting remarks that the attractive ability of amber was destroyed when some material is interposed between it and its victim whereas that of loadstone was unaffected under similar conditions, and that, when loadstone attracted iron, the iron was also attracted by the loadstone whereas no such reciprocal action was observed with amber. Finally, he made the interesting observations that loadstone attracted towards its ends (poles) whereas no such 'polarity' was observed in amber and that, whereas friction increased (or was necessary for) the attractive powers of amber, no similar effect was observed with loadstone except, of course, by cleaning the poles. Cardan's work was followed by what is usually regarded as the most significant early piece of research on electricity and magnetism (mainly on magnetism) by William Gilbert (1544–1603). Gilbert, who was born in Colchester, England, and educated at Cambridge University, published the influential book De Magnete (On the magnet) in 1600.

In this book, he showed himself to be an exhaustive experimenter, finding a large number of materials which possessed the properties of amber, including the already known materials, jet (a kind of compacted coal) and diamond and named 'electrics' by him. He also gave a more complete list of which materials would be attracted by his electrics. In the process of these experiments, Gilbert invented what was, in a sense, the first electroscope (called, by Gilbert, the versorium); a fine needle of metal (or certain other substances) delicately balanced on a sharp point and which could rotate freely upon it. A rubbed material could then be brought close to it and even very small attractions between material and needle could be observed and recorded. Gilbert tested many hypotheses; in particular, he showed that heating amber is not enough to give it its attractive property. It is only when this heating comes from rubbing that this property is found. He re-examined the effect that one can shield the force of an electric but not that of a loadstone, coming to the same conclusion as Cardan. He also demonstrated that electric and magnetic attraction between two bodies decreased as the distance between them increased and was aware of the fact that, on making contact with an electric, an attracted body would soon fall away.

In his experimental work on magnetism, he described what would now be called the magnetic field by displaying the familiar patterns around a magnet using iron filings. (Descartes' suggestion of a fluid vortex theory of magnetism could also be interpreted as containing the germ of the magnetic field concept.) Gilbert regarded the Earth as a large magnet (linking it to the rotation of the Earth) and was thus able to explain the directional properties of the compass. He was, in fact, responsible for the terms north and south poles. He was aware of the magnetic dip of the Earth (probably first studied in detail by Robert Norman in 1581) and the difference between the true and the magnetic poles. Gilbert, although he framed several hypotheses in his famous book, relied on the experimental method and, like his contemporaries, Galileo and Francis Bacon, was one of the original exponents of experimental philosophy.

EXPERIMENTAL DEVELOPMENT OF ELECTRICITY

In 1629, Niccolo Cabeo wrote an important work that listed even more materials that could be classified as electrics and also observed repulsion of electrics (but interpreted it as a 'bouncing-off' effect). Other later Italian workers observed and interpreted correctly the mutual attraction effect for electrics and experimented to see if the attractive properties of amber were due to the surrounding air by attempting the experiment in vacuo. Their attempt was inconclusive but a successful answer was provided by Robert Boyle in England in the 1670s (following the development of a better vacuum pump) when he showed the effect to be still present. In the early eighteenth century, a series of experiments was carried out by Francis Hauksbee in England. He observed the effects of electric attraction on suspended threads and revealed the electric field in a similar way to Gilbert's unveiling of the magnetic field. He also 'felt' the 'electric wind' caused by a stream of charged particles emanating from a rubbed electric. In 1729, Steven Gray observed the feature of 'charging by conduction' and, a little later, that of 'charging by influence'. He was aware of the difference between these in that the latter, unlike the former, even for an insulated body, was present only as long as the influencing body was. Another crucial difference would, of course, be noted later. He also had the concept of electric current. Using a coarse thread as a 'wire', he was able to charge one end and record it at the other end even though the thread was of significant length. He was also aware that this charge transfer would be improved if the thread was wet and also that some materials did not transmit charge. Thus, he had identified and distinguished between 'conductors' and 'insulators'.

In a letter to the Royal Society, published in 1734, the Frenchman, Charles du Fay, extended the list of electrics to essentially anything that could be rubbed (metals excluded) and also described the conduction of 'electricity' along threads. Further, he identified the repulsion of an object from the electric to which it had originally been attracted, that this repulsion from the original electric was retained until the object was 'discharged' and that it may, before discharge, be attracted to other electrics. He was the first to formulate the idea of two types of electricities that he called 'vitreous' and 'resinous' (largely dependent on the material being rubbed) and which could neutralize each other and where one (either) type could be created by removing some of the other type. He also developed the concept of 'amount' of electricity in the sense that equal amounts of each type of electricity would exactly neutralize each other. Later, of course, it was discovered that the type of electricity depended on both the rubbed and the rubbing materials. In the middle of the eighteenth century, Benjamin Franklyn recognized only one type of electricity but that it could occur in two forms which he called 'plus' or 'positive' and 'minus' or 'negative' and which corresponded to the vitreous and resinous electricity of du Fay, respectively. Here, as with du Fay, is the suggestion of the well-known law of charge conservation. Franklyn is also known for his experiments with lightning conductors. The eighteenth century also witnessed the development of superior sources of electricity for experimentation, such as the Leyden jar, which gave a heavy and short-lived discharge, and the voltaic pile, which gave a more continuous flow of electricity.

Perhaps the only truly quantitative results about electricity and magnetism developed in this period were the so-called Coulomb laws regarding the force between two charges or between two magnetic poles. Using a torsion balance, the French Scientist, Charles Augustin Coulomb, showed in 1785 that the magnitude of the force between two fixed charges or two fixed magnetic poles varies as the inverse square of their distance apart (including the sign in the force for the cases when the charges or poles are of the same or opposite type). The ability to measure, with accuracy, quantity of charge was not well developed in Coulomb's day, but later work confirmed that the force between two charges of whatever sign is, in magnitude, also proportional to the product of the quantities of the charges involved. Thus, one has the familiar Coulomb law. This result, for charges, was also known earlier to some extent to Daniel Bernoulli in ca 1760, John Robison in Edinburgh in 1769 and Henry Cavendish in 1771. An indirect suggestion that the inverse square law was the appropriate law of attraction was given by Joseph Priestly in 1767 with an appeal to a result in potential theory. Coulomb's law for magnetic poles was probably first given by John Michell in Cambridge in 1750.

In summary, one can say that, by the beginning of the nineteenth century, physicists (or natural philosophers as they may have preferred to have been known) were aware of magnets, electrics and some of their properties. They had elementary 'batteries' and had sent electric currents to significant distances. They knew of insulators and conductors and had reduced electric charge to two types. With Coulomb and others, they had a quantitative law about the mutual effects of charges and of magnetic poles and some concept of electric and magnetic fields. Perhaps, more importantly, they had learned (from, among others, Gilbert, Galileo and Bacon) modern experimental physics in the sense that they realized the importance of performing experiments and recording the results carefully. However, the study of electrics and magnetic materials had been largely kept separate in spite of the similarities between them. It was at the beginning of the nineteenth century that their interrelationships were discovered and that they were merged into a single discipline.

ELECTRICITY AND MAGNETISM

In 1820, the Danish scientist, Hans Christian Oersted, apparently inspired by the effect of a thunderstorm on a compass needle and suspecting lightning as the culprit, performed an experiment to show that, if a compass needle is positioned above or below a current-carrying wire, it attempts to set with its long axis perpendicular to the wire. This was perhaps the first demonstration of a direct link between electricity and magnetism, that is, between the electricity in the current and the magnetic effects on the compass needle, rather than the revealing of another similarity between electric and magnetic effects. (It is appropriate at this point to stress the importance for this, and other similar experiments, of the discovery of the voltaic pile as a provider of a continuous source of current.) Very soon after, the French scientists, Jean-Baptiste Biot and Félix Savart, in their attempts to find more qualitative results showed experimentally that the actual force on a magnetic pole due to the current in a long straight wire was proportional to the current in the wire, inversely proportional to the perpendicular distance from the point to the wire and acted in the direction perpendicular to the plane containing the wire and the point. They then investigated the problem theoretically and found the familiar differential form of the law. In this differential form, where the force due to a current-carrying 'element' of wire is written down, the familiar inverse square law is again encountered which then integrates out to give the above-mentioned law. Also, in 1820, André-Marie Ampère undertook further mathematical study of this phenomenon, showing how two parallel current-carrying wires attract or repel each other, depending on whether the currents are in the same or opposite direction. He also wrote down an expression for this force for more general kinds of currents. Since this force is essentially the magnetic field, Ampère's work gives a very general expression for the magnetic field generated by a current. His work was of crucial importance for what was to follow several decades later (and Maxwell was to refer to him as the 'Newton of electricity'). Thus, not only could a current-carrying wire affect a magnet, as Oersted had showed, but it could also affect another current-carrying wire. But if a current-carrying wire (i.e. charges in motion) could create a magnetic field, could a magnetic field create a current?

In 1831, in a paper read to the Royal Society, the English scientist, Michael Faraday, described a most remarkable experiment (see Faraday 1839–1855). He wound two separate insulated coils of wire onto an iron ring with one coil attached to a galvanometer and the other to a battery. When he switched the battery current on, the galvanometer needle flickered, as it did again when he switched it off. In the intervening stationary period, and although current still flowed in the second (battery) wire, the galvanometer recorded no current in the first wire. Thus, he demonstrated that a changing magnetic field, caused by the current starting up or closing down in the second wire, created a current in the first (galvanometer) wire. He also showed that the same current creation could be achieved by simply moving a magnet (again creating a changing magnetic field) inside a wire coil which was attached to a galvanometer. No such effect on the galvanometer was recorded if the magnet was nearby but stationary. Thus, a changing magnetic field was capable of creating a current in a conductor. (This phenomenon was also apparently known to the American, Joseph Henry, but he failed to publish before Faraday.) This remarkable result of Faraday was to be put to many practical uses including the electric motor and the dynamo for generating electricity.

Faraday's ideas rested not on mathematics, for Faraday had no mathematical training, but on his concept of lines of force. (However, Faraday's thinking was, in its nature, mathematical (geometrical) and was found to be rather useful in the later developments by the mathematical physicists.) Thus, Faraday imagined such lines emanating out from electric charges (electric lines of force) or from magnetic poles (magnetic lines of force). Indeed, such visual concepts are still used in understanding the theory of electricity and magnetism, even if the mathematics that was ultimately to describe this theory had no need of them. Faraday imagined that the space in which a magnetic field operated was in a special state, called electrotonic and his lines of force are particularly instructive, geometrically, in understanding, for example, the well-known patterns observed in the vicinity of a magnet (as observed by Gilbert) by covering it with a horizontal piece of paper and sprinkling iron filings on the paper. Lines of force are also useful in many applications of electromagnetic theory; for example, the current observed in Faraday's experiment can also be created by moving the galvanometer wire in the vicinity of a stationary magnet. Thus, the consequences of Faraday's experiment really depend only on the relative motion of the magnet and the wire. So one can picture the Faraday experiment by imagining that the current in the wire is created by the changes of the lines of magnetic force across the galvanometer wire and this picture is faithful to the fact that only relative velocities are involved. Thus, the experiments of Oersted, Ampère, Faraday and others display the interrelations between currents (moving charges) and magnetic fields. (However, as remarked by Einstein in his original paper on special relativity, although the effect is experimentally symmetrical, in the above sense of relative velocity, the conventional way of describing it is not. For when the magnet moves, one seeks an explanation for the induced current in the changing magnetic field in the vicinity of the wire whereas if the magnet is at rest one finds the reason for the current in the forces acting on the moving charges in the wire within the magnetic field created by the magnet.)

THE PERIOD OF MAXWELL

4In the middle of the nineteenth century, one could, albeit rather simplistically, regard research into mathematical electromagnetic theory as divided into two camps. The first of these, mostly in continental Europe and following Ampère, Weber and Neumann, were Newtonians in that they formulated theories that were based on action at a distance (as Newtonian gravity was). In Great Britain, Faraday, Kelvin and Maxwell preferred what might be called field theories in which electric and magnetic effects were either transmitted by some medium (an ether) or by Faraday's lines of force described earlier. The concept of an ether had been given extra impetus from the demonstration of the wave-like character of light by Huygens, Fresnel and Young. Faraday even suggested that light was a kind of undulatory motion in his lines of force and supported this view by his discovery of the so-called Faraday effect in which the plane of polarization of a light beam was rotated by a nearby magnetic field (or current flowing in a wire around the light beam). Thus, he found an effect linking electricity, magnetism and light.

Maxwell's first real step towards his electromagnetic theory was in the paper On Faraday's lines of force (in two parts in 1855–1856; Maxwell 1864). In this paper, he proceeded by analogy, as was not atypical of Maxwell, basing his techniques on those of Faraday. He introduced into space a hypothetical incompressible fluid, the flow lines of which could be taken as a geometrical representation of the electric or magnetic field, or the current flow. For example, in the case of electric fields, the sources of the flow lines were positive charges and the sinks were negative charges. There was difficulty in this geometrical approach in incorporating the existence of an interaction between the electric and magnetic fields and the current. But he was able to provide a mathematical description of several of Faraday's ideas. In his next paper On physical lines of force(in four parts 1861–1862), he introduced his famous 'vortices' or 'cells'.

This was an attempt to incorporate the known 'rotational' properties of the magnetic field into his model. He envisaged these vortices (perhaps one can think of them as very small and spherical) occupying space (even the vacuum) and to be of small but non-zero mass, and elastic. However, their properties could change depending on the medium they occupied. These vortices were separated by particles (which one can think of as small balls) which acted as idle wheels. Thus, the idle wheels (particles), rotating without translational movement, would, from the rolling condition, gear with the vortices and enable them to rotate in the same sense. However, any translational movement of the particles would generate a differential rotation of the vortices.

The rotating vortices are to be thought of as representing the magnetic field, whose magnitude is essentially the rotational speed of the vortex and whose direction is determined by the vortex's axis and sense of rotation. The movement of the balls is interpreted as electric current. With these interpretations appropriately made, Maxwell was able to derive Ampère's circuital law and to provide an explanation of Faraday's law of induction. In Maxwell's model, the balls were free to move in a conductor but not in a dielectric (an insulator). In the third part of his paper, and using his elasticity assumption, he was able to explain the familiar laws of attraction between charges. Further, he was also able to suggest that, in a dielectric medium, an electric field can slightly displace the electrical balls and that, during the initial phase of this change, when the electric field is changing, this movement of the balls gives rise to what is essentially a transient electric current. This is Maxwell's displacement current and his model showed that it was intimately connected to the rate of change of the electric field. In the same part of this paper, and as will be dealt with in more detail later, he pointed out how his model suggested that changes in these electromagnetic fields would be propagated through this elastic medium at a finite, if large, speed and which he could calculate approximately. The value he obtained was rather close to the experimentally accepted value of the speed of light in the supposed 'luminiferous' ether in which light was assumed to travel. He concluded that light consisted of transverse vibrations in the (electromagnetic) ether and suggested identifying the latter with the luminiferous ether. In the fourth part of his paper, he offered an explanation for the Faraday effect mentioned earlier.

Maxwell's introduction of the displacement current modified Ampère's circuital law because now there was an extra current to be included in addition to the conduction current. Now the equations of electromagnetic theory, including the equation of charge conservation, were, prior to the introduction of this extra current by Maxwell, internally inconsistent, at least for the case of a charging capacitor in open circuit. Maxwell's inclusion of the displacement current in Ampère's law corrected this deficiency. This term plays the role, for example, of a current in the gap between the plates of a capacitor during charging.

In the key paper of this series and carrying the title A dynamical theory of the electromagnetic field(see Niven 1965), Maxwell attempted to remove from his theory some of the appeals to models and analogies and to lay out a coherent mathematical theory of electromagnetism. He was also able to derive from his equations an electromagnetic theory of light. He had built upon the earlier physical and mathematical developments of Ampère, Faraday, Kelvin, Weber, Green and Stokes and had derived the equations describing the behaviour of the electromagnetic interaction. Maxwell's equations, as they now stood (and in modern notation following the later simplifying work of Heaviside (who removed the potentials in Maxwell's original expressions so that the equations were written entirely in terms of the electric and magnetic fields and their sources), Hertz and Lorentz), are given by

$$\nabla \cdot \boldsymbol{D} = 4\pi\rho,$$
$$\nabla \cdot \boldsymbol{B} = 0,(5$$
$$\nabla \times \boldsymbol{H} = (4\pi/c)\boldsymbol{j} + 1/c\, \partial \boldsymbol{D}/\partial t,$$
$$\nabla \times \boldsymbol{E} = -1/c\, \partial \boldsymbol{B}/\partial t,($$

where D is the electric (displacement) field; E is the electric field; B is the magnetic (induction) field; H is the magnetic field; j is the conduction current density vector field; ñ is the charge density; and " is the usual differential operator. The quantity c will be described later. These equations would normally be augmented by the so-called constituent relations relating E and D, Hand B and j and E but they will not be needed here except to say that, in vacuo, E=D, H=B, j=0 and ñ=0. Equation (5.1) arises from Coulomb's law for the force between electric charges, while (5.2) is the statement of the absence of free magnetic poles. Equation (5.3) is Maxwell's corrected form of Ampère's circuital law relating the magnetic field to the current and where the conduction current j has been augmented by the Maxwell displacement current$1/4\pi\, \partial \boldsymbol{D}/\partial t$. Thus, the set of equations above was now consistent with the law of charge conservation,$\partial \rho/\partial t + \nabla \cdot \boldsymbol{j} = 0$. Equation (5.4) is Faraday's induction law. The quantity c can be calculated from the electromagnetic properties of the medium under consideration. (In this respect, the work of the German scientists, Wilhelm Weber and Carl Friedrich Gauss, must be mentioned since it was they who were the major forces in the determination of the system of units used in electromagnetic theory and which is intimately related to this problem.) In vacuo, where E=D and H=B, c turned out to be a constant rather close to the value of the speed of light accepted at the time. In this case, a rather well-known and straightforward manipulation of Maxwell's equations, using some standard vector analysis, then shows that E and B satisfy the relations

$$\nabla^2 E = 1/c^2\partial^2 E/\partial t^2, \qquad \nabla^2 B = 1/c^2\partial^2 B/\partial t^2,$$

where " is the usual Laplacian operator. Thus, the electric and magnetic fields satisfy the familiar wave equation with propagation speed c and which is now interpreted as the speed of light. This was the main supporting evidence that light was an electromagnetic phenomenon. In this sense, Maxwell had succeeded in theoretically unifying electricity, magnetism and light and pressed his claim for the need of only one ether with respect to which c was the speed of light. Maxwell gathered together his work and the work of others into his great mathematical work A treatise on electricity and magnetism (1873) and which was mostly written on his estate at Glenlair. In a series of experiments in 1887–1888, sadly 9 years after the death of Maxwell, the German physicist, Heinrich Hertz, experimentally detected Maxwell's electromagnetic waves (essentially radio waves) and confirmed their speed to be of the same order of magnitude as the speed of light. (In fact the English physicist, Hughes, had similarly detected such waves some 7 years earlier than Hertz but became discouraged when it was suggested that his results may be explained by electromagnetic induction, and Hertz published first.) The celebrated experiment of Hertz recorded results that were consistent with a Maxwell-type theory in which effects were propagated with a finite velocity and which were not consistent with 'action at a distance' theories, in which effects were propagated instantaneously. It may be said that, from the date of Hertz's success, Maxwell's theory was essentially fully accepted.

FROM MAXWELL TO EINSTEIN

Towards the end of the nineteenth century, theoretical physics was dominated by the mechanics of Newton and the electromagnetic theory of Maxwell. In Newton's theory, one has the conc(pt of absolute time and a system of preferred observers, the inertial observers. Such inertial observers, or inertial frames of reference, arise from the fact that Newton's theory is based on the concept of force and essentially claims to be able to distinguish between what might be called 'real' forces and 'inertial' (also called 'accelerative') forces. Then Newton's second law, which equates the real force on a particle to the product of its mass and its acceleration, holds only in such privileged inertial frames. Each inertial observer is, then, a representative of Newton's absolute space and any non-inertial observer will experience accelerative forces as a kind of reaction on him due to his acceleration with respect to (any of) these absolute spaces. The inertial observers, regarded in the usual way as coordinate systems, are connected, mathematically, by the so-called Galileantransformations and Newton's relativity principle is the statement that these inertial observers are indistinguishable one from another by mechanical experiments (Galilean invariance). However, Maxwell's theory now introduces an ether and the crucial constant c, the speed of light in the rest frame of the ether. Thus, if one assumes the existence of an inertial observer who is at rest with respect to this ether, this observer measures the speed of light, independently of the latter's direction, to be precisely this number c. However, other inertial observers in relative motion with respect to this ether observer, will clearly not record such a convenient measurement owing to the addition laws for velocity derived from the Galilean transformations in Newtonian mechanics. Thus, an electromagnetic experiment could distinguish one inertial frame from another and the Galilean invariance of Maxwell's theory is lost.

In this period, one of the first duties was to find the relative speed v of Earth and ether and the most celebrated experiment in this direction was performed, following Maxwell's suggestion, by Michelson and Morley in 1887 (following an earlier flawed one by Michelson in 1881). In this experiment, the quantity being measured, that is the above relative speed, occurred in the equations in the combination (v/c) and was thus a 'second-order' effect. It was considered impossible at that time to measure such small quantities directly, but Michelson's interferometer, taking advantage of wave interference, was capable of the task. The conclusion was that, assuming the prevailing Newtonian theory and Galilean transformations, no motion of the Earth through the ether was detected. Other similar experiments were also performed but with the same general result. Many explanations were forthcoming to explain these so-called 'null' results, of which perhaps the most famous were the contraction hypotheses of Fitzgerald and Lorentz. These hypotheses actually envisaged that a rigid rod moving with respect to the ether would suffer a physical shrinking.

The history of special relativity (for more details, see Whittaker 1951) may be taken to begin with Poincaré in 1899 when he claimed that optical experiments really depended only on the relative velocity of the bodies in question and that absolute motion in this sense was undetectable. Thus, Poincaré in 1904 and Einstein in 1905 announced the principle of relativity stating that noexperiment could distinguish one inertial frame from another and hence that all physical laws should reflect this invariance. Thus, it extended the Newtonian principle of relativity from mechanical to all experiments. The true understanding of what is now called special relativity originates with Einstein's paper of 1905. Although it is not entirely clear whether Einstein was aware of the Michelson–Morley experiment at the time, it appears that he probably was. In any case, Einstein had no need of an ether and accordingly rejected it. He argued, roughly speaking, that one should keep the inertial frames of Newtonian theory but should include, in addition, this much stronger principle of relativity, together with the assumption that each inertial observer would measure exactly the same speed of light, independently of its direction with respect to that observer. Clearly, one cannot retain the Galilean transformations, since they are obviously inconsistent with such an assumption. Einstein then proceeded to calculate the (consistent) transformations between inertial frames needed to replace these Galilean transformations and gave the first real physical derivation of the (Lorentz) transformations (which had, in fact, been used earlier by Voigt and Lorentz and named after Lorentz by Poincaré). It followed from this that Newton's absolute time had to be abandoned as inconsistent with these transformations (but absolute space was essentially as in Newton's theory). Einstein was also able to show that the contraction hypotheses of Fitzgerald and Lorentz arose naturally in his theory and were of a kinematic rather than a physical nature. He further showed that the preservation of time by the Galilean transformations in Newtonian theory was replaced, in his theory, by an expression describing the 'relative' nature of time for inertial observers.

Following an earlier and incomplete attempt by Lorentz (of which Einstein was apparently unaware), Einstein was able to use the assumption that Maxwell's equations are invariant in form under the Lorentz transformations (as required by the principle of relativity) to find the relations between the electric and magnetic fields in two inertial frames. It turned out that the electric field for one inertial observer is a particular combination of the electric and magnetic fields of the other observer (the precise nature of which depends on the kinematical relationship between the two observers) and similarly for the magnetic field. It follows that the electric and magnetic fields do not, in this sense, have a separate existence but rather are observer-dependent manifestations of a single electromagnetic field. Einstein, in fact, in his 1905 paper referred to the strange asymmetry in the physical description of the important experiment of Faraday (as mentioned in §4) as being at variance with the experimental symmetry.

This problem arises as a consequence of its being described by a single observer in terms of his electric and magnetic fields and the dependence of these fields on the observer. Thus, Maxwell's work, together with Einstein's special relativity theory, constituted a unification of the electric and magnetic fields into a more meaningful electromagnetic field. Of course Maxwell's equations, as written above, still refer to the electric and magnetic fields separately. The complete mathematical unification of these fields came a little later, from the beautiful work of Hermann Minkowski, when he introduced the Maxwell tensor (or Maxwell bivector) into the theory in 1908. This mathematical object has six degrees of freedom, representing the three components of each of the electric and magnetic fields and which, owing to its tensor nature, could be regarded as being, in a well-defined sense, independent of the observer. However, a particular observer may still retrieve his personal electric and magnetic fields from it by a simple procedure. Maxwell's equations may then be rewritten in a simple and elegant way as differential equations involving this tensor.

Maxwell's electromagnetic theory and Einstein's special theory of relativity have withstood the test of time and stand as remarkable tributes to these scientists. The theory of electricity and magnetism is one of the few areas of physical enquiry that did not attract Newton's attention. It differed in many ways from Newtonian philosophy. For example, the magnetic field due to a current element which is experienced at a certain point does not act along the straight line joining the element and point but rather at right angles to this line. Thus, whereas classical ideas, embodied in the law of inertia, lead to the identification of straight lines in the (assumed) Euclidean structure of space, electromagnetic th ɜory identifies right angles in this structure.

Maxwell will be remembered for his many contributions to theoretical physics and his work on electricity and magnetism and the consequent Maxwell equations are probably the most important addition to theoretical science in the nineteenth century.

Maxwell's Theory of Electromagnetism James Clerk Maxwell (first mentioned in Lesson 21) is one of the great tragedies of physics. Ï% Although he contributed a huge amount of work to physics, he died in 1879 at the age of 48. Ë% Many scientists believe he was the most important physicist of the 19th century. Ë% Some of his papers found at the time of his death were later interpreted to show that he was working on ideas that would have eventually led him to the Special Theory of Relativity, twenty years before Einstein! Maxwell's greatest accomplishment was his Theory of Electromagnetism. Ï% Oersted had found that we can create a magnetic field by having current flow through a wire. Ë% The strength of the magnetic field is predicted using Ampere's Law (Lesson 18). Ï% Moving a wire through a magnetic field induces a current to flow through the wire. Ë% The amount of current induced is predicted using Faraday's Law of Induction (Lesson 20). Ï% Maxwell brought these ideas together while doing experiments with capacitors. Ë% A capacitor is made up of two conductors that are not touching, each with the same amount of oppositely signed charge. Ë% It was while using capacitors that he came up with an experiment that showed that a changing electric field resulted in a changing magnetic field which created a changing electric field which... you get the idea.

MAXWELL MADE A SERIES OF PREDICTIONS BASED ON HIS RESEARCH:

1. EMR is produced whenever a charge accelerates. This is because the accelerating charge will produce a changing electric field, the start of the pattern of interrelated fields discussed already. It also means that

the accelerating charge must be losing energy since it is emitting EMR (a form of energy).

2. The frequency that a charge jiggles back and forth with will exactly match the frequency of the EMR produced. This makes sense, since the accelerating charge is what originally causes the fields that make up the EMR.
3. All EMR travels at the speed of light, "c" Maxwell actually predicted a speed slightly higher than the current accepted value of 3.00e8 m/s, but give the guy a break! This was over 140 years ago!
4. The electric and magnetic fields are perpendicular to each other, and perpendicular to the direction the wave is traveling. Picture the electric and magnetic fields as intertwined transverse waves. Since they create each other forever, we sometimes refer to EMR as a self propagating wave.
5. EMR will do all the things any transverse wave should do. This includes ideas we will look at in later lessons like refraction, interference, and polarization. He also showed that we could use the universal wave equation c = fë for EMR

4

Faraday's law on Electromagnetism

AND ITS PROOF

Faraday's law of induction is a basic law of electromagnetism predicting how a magnetic field will interact with an electric circuit to produce an electromotive force (EMF)—a phenomenon called electromagnetic induction. It is the fundamental operating principle of transformers, inductors, and many types of electrical motors, generators and solenoids. The Maxwell–Faraday equation is a generalization of Faraday's law, and is listed as one of Maxwell's equations.

HISTORY

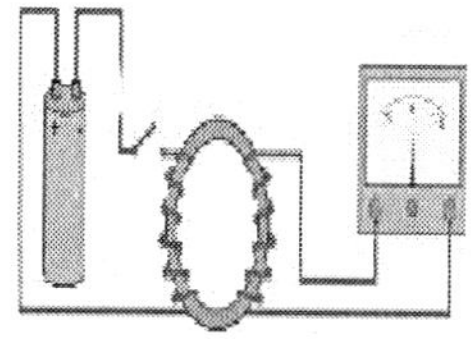

A diagram of Faraday's iron ring apparatus. The changing magnetic flux of the left coil induces a current in the right coil.

Faraday's disk, the first electric generator, a type of homopolar generator.

Electromagnetic induction was discovered independently by Michael Faraday in 1831 and Joseph Henry in 1832. Faraday was the first to publish the results of his experiments. In Faraday's first experimental demonstration of electromagnetic induction (August 29, 1831), he wrapped two wires around opposite sides of an iron ring (torus) (an arrangement similar to a modern toroidal transformer). Based on his assessment of recently discovered properties of electromagnets, he expected that when current started to flow in one wire, a sort of wave would travel through the ring and cause some electrical effect on the opposite side. He plugged one wire into a galvanometer, and watched it as he connected the other wire to a battery. Indeed, he saw a transient current (which he called a "wave of electricity") when he connected the wire to the battery, and another when he disconnected it. This induction was due to the change in magnetic flux that occurred when the battery was connected and disconnected. Within two months, Faraday had found several other manifestations of electromagnetic induction. For example, he saw transient currents when he quickly slid a bar magnet in and out of a coil of wires, and he generated a steady (DC) current by rotating a copper disk near the bar magnet with a sliding electrical lead ("Faraday's disk").

Michael Faraday explained electromagnetic induction using a concept he called lines of force. However, scientists at the time widely rejected his theoretical ideas, mainly because they were not formulated mathematically. An exception was James Clerk Maxwell, who in 1861-2 used Faraday's ideas as the basis of his quantitative electromagnetic theory. In Maxwell's papers, the time-varying aspect of electromagnetic induction is expressed as a differential equation which Oliver Heaviside referred to as Faraday's law even though it is different from the original version of Faraday's law, and does not describe motional EMF. Heaviside's version (see Maxwell–Faraday equation below) is the form recognized today in the group of equations known as Maxwell's equations.

FARADAY'S LAW AND RELATIVITY

TWO PHENOMENA

Faraday's law is a single equation describing two different phenomena: the motional EMF generated by a magnetic force on a moving wire (see Lorentz force), and the transformer EMF generated by an electric force due to a changing magnetic field (due to the Maxwell–Faraday equation).

James Clerk Maxwell drew attention to this fact in his 1861 paper On Physical Lines of Force. In the latter half of Part II of that paper, Maxwell gives a separate physical explanation for each of the two phenomena.

A reference to these two aspects of electromagnetic induction is made in some modern textbooks. As Richard Feynman states:

So the "flux rule" that the emf in a circuit is equal to the rate of change of the magnetic flux through the circuit applies whether the flux changes because the field changes or because the circuit moves (or both)...

Yet in our explanation of the rule we have used two completely distinct laws for the two cases $- v \times B$ for "circuit moves" and $'' \times E = \rceil''_t B$ for "field changes".

We know of no other place in physics where such a simple and accurate general principle requires for its real understanding an analysis in terms of two different phenomena.

Richard P. Feynman, The Feynman Lectures on Physics

EINSTEIN'S VIEW

Reflection on this apparent dichotomy was one of the principal paths that led Einstein to develop special relativity:

It is known that Maxwell's electrodynamics—as usually understood at the present time—when applied to moving bodies, leads to asymmetries which do not appear to be inherent in the phenomena. Take, for example, the reciprocal electrodynamic action of a magnet and a conductor.

The observable phenomenon here depends only on the relative motion of the conductor and the magnet, whereas the customary view draws a sharp distinction between the two cases in which either the one or the other of these bodies is in motion. For if the magnet is in motion and the conductor at rest, there arises in the neighbourhood of the magnet an electric field with a certain definite energy, producing a current at the places where parts of the conductor are situated.

But if the magnet is stationary and the conductor in motion, no electric field arises in the neighbourhood of the magnet. In the conductor, however, we find an electromotive force, to which in itself there is no corresponding energy, but which gives rise—assuming equality of relative motion in the two cases discussed—to electric currents of the same path and intensity as those produced by the electric forces in the former case.

Examples of this sort, together with unsuccessful attempts to discover any motion of the earth relative to the "light medium," suggest that the phenomena of electrodynamics as well as of mechanics possess no properties corresponding to the idea of absolute rest.

FARADAY'S VIEW ON ELECTROMAGNETIC INDUCTION

In 1831, Michael Faraday, an English physicist gave one of the most basic laws of electromagnetism called Faraday's law of electromagnetic induction. This law explains the working principle of most of the electrical motors, generators, electrical transformers and inductors. This law shows the relationship between electric circuit and magnetic field. Faraday performs an experiment with a magnet and a coil. During his experiment, he found how emf is induced in the coil when flux linked with it changes.

FARADAY'S EXPERIMENT

RELATIONSHIP BETWEEN INDUCED EMF AND FLUX In this experiment, Faraday takes a magnet and a coil and connects a galvanometer across the coil. At starting, the magnet is at rest, so there is no deflection in the galvanometer i.e needle of galvanometer is at the center or zero position. When the magnet is moved towards the coil, the needle of galvanometer deflects in one direction. When the magnet is held stationary at that position, the needle of galvanometer returns back to zero position. Now when the magnet is moved away from the coil, there is some deflection in the needle but in opposite direction and again when the magnet becomes stationary, at that point with respect to coil, the needle of the galvanometer returns back to the zero position. Similarly, if magnet is held stationary and the coil is moved away and towards the magnet, the galvanometer shows deflection in similar manner. It is also seen that, the faster the change in the magnetic field, the greater will be the induced emf or voltage in the coil.

FARADAY'S FIRST LAW

Any change in the magnetic field of a coil of wire will cause an emf to be induced in the coil. This emf induced is called induced emf and if the conductor circuit is closed, the current will also circulate through the circuit and this current is called induced current. Method to change magnetic field:

1. By moving a magnet towards or away from the coil
2. By moving the coil into or out of the magnetic field.
3. By changing the area of a coil placed in the magnetic field
4. By rotating the coil relative to the magnet.

Thus, according to Faraday, Whenever a conductor is placed in a varying magnetic field an EMF gets induced across the conductor (called as induced emf), and if the conductor is a closed circuit then induced current flows through it.

Magnetic field can be varied by various methods - 1. By moving magnet, 2.By moving the coil, 3. By rotating the coil relative to magnetic field

FARADAY'S SECOND LAW

It states that the magnitude of emf induced in the coil is equal to the rate of change of flux that linkages with the coil. The flux linkage of the coil is the product of number of turns in the coil and flux associated with the coil.

- By increasing the number of turns in the coil i.e N, from the formulae derived above it is easily seen that if number of turns in a coil is increased, the induced emf also gets increased.
- By increasing magnetic field strength i.e B surrounding the coil-Mathematically, if magnetic fieldincreases, flux increases and if flux increases emf induced will also get increased. Theoretically, if

the coil is passed through a stronger magnetic field, there will be more lines of force for coil to cut and hence there will be more emf induced.

- By increasing the speed of the relative motion between the coil and the magnet - If the relative speed between the coil and magnet is increased from its previous value, the coil will cut the lines of flux at a faster rate, so more induced emf would be produced.

Thus, Faraday's second law of electromagnetic induction states that, the magnitude of induced emf is equal to the rate of change of flux linkages with the coil. The flux linkages is the product of number of turns and the flux associated with the coil.

FORMULA OF FARADAY'S LAW

Consider the conductor is moving in magnetic field, thenflux linkage with the coil at initial position of the conductor = $NÖ_1$ (Wb) (N is speed of the motor and Ö is flux)

flux linkage with the coil at final position of the conductor = $NÖ_2$ (Wb)change in the flux linkage from initial to final = $N(Ö_1 - Ö_2)$ let $Ö_1 - Ö_2$ = Ötherefore, change in the flux linkage = NÖand, rate of change in the flux linkage = NÖ/ttaking the derivative of RHSrate of change of flux linkages = N (dÖ/dt)

According to Faraday's law of electromagnetic induction, rate of change of flux linkages is equal to the induced emf

So, E = N (dÖ/dt) (volts)

PHENOMENON OF MUTUAL INDUCTION

Alternating current flowing in a coil produces alternating magnetic field around it. When two or more coils are magnetically linked to each other, then an alternating current flowing through one coil causes an induced emf across the other linked coils. This phenomenon is called as mutual induction.

LENZ'S LAW

Lenz's law of electromagnetic induction states that, when an emf is induced according to Faraday's law, the polarity (direction) of that induced emf is such that it opposes the cause of its production.

THUS, CONSIDERING LENZ'S LAWE = -N (DÖ/DT) (VOLTS)

The negative sign shows that, the direction of the induced emf and the direction of change in magnetic fields have opposite signs.

APPLICATIONS OF FARADAY'S LAW

Faraday law is one of the most basic and important laws of electromagnetism. This law finds its application in most of the electrical machines, industries and medical field etc.

- Electrical Transformers work on Faraday's law
- The basic working principle of electrical generator is Faraday's law of mutual induction.

- The Induction cooker, is a most fastest way of cooking. It also works on principle of mutual induction. When current flows through the coil of copper wire placed below a cooking container, it produces a changing magnetic field. This alternating or changing magnetic field induces an emf and hence the current in the conductive container, and we know that flow of current always produces heat in it.
- Electromagnetic Flow Meter is used to measure velocity of certain fluids. When a magnetic fieldis applied to electrically insulated pipe in which conducting fluids are flowing, then according to Faraday's law, an electromotive force is induced in it. This induced emf is proportional to velocity of fluid flowing.
- Form the bases of Electromagnetic theory, Faraday's idea of lines of force is used in well known Maxwell's equations. According to Faraday's law, change in magnetic field gives rise to change in electric field and the converse of this is used in Maxwell's equations.
- It is also used in musical instruments like electric guitar, electric violin etc.

If the wire is then wound into a coil, the magnetic field is greatly intensified producing a static magnetic field around itself forming the shape of a bar magnet giving a distinct North and South pole.

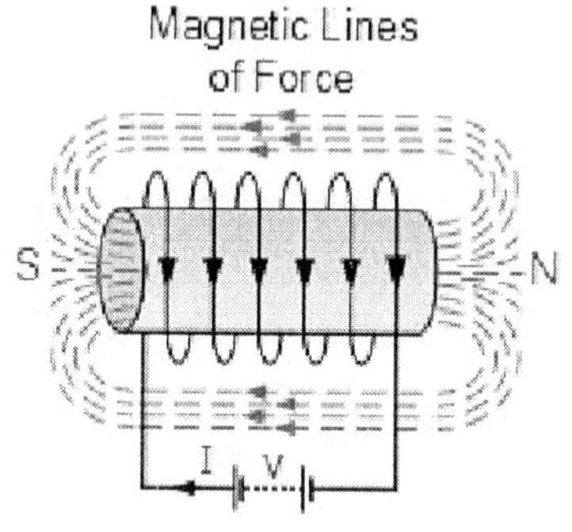

AIR-CORE HOLLOW COIL

The magnetic flux developed around the coil being proportional to the amount of current flowing in the coils windings as shown. If additional layers of wire are wound upon the same coil with the same current flowing through them, the static magnetic field strength would be increased.

Therefore, the magnetic field strength of a coil is determined by the ampere turns of the coil. With more turns of wire within the coil, the greater the strength of the static magnetic field around it.

But what if we reversed this idea by disconnecting the electrical current from the coil and instead of a hollow core we placed a bar magnet inside the core of the coil of wire. By moving this bar magnet "in" and "out" of the coil a current would be induced into the coil by the physical movement of the magnetic flux inside it.

Likewise, if we kept the bar magnet stationary and moved the coil back and forth within the magnetic field an electric current would be induced in the coil. Then by either moving the wire or changing the magnetic field we can induce a voltage and current within the coil and this process is known as Electromagnetic Induction and is the basic principal of operation of transformers, motors and generators.

Electromagnetic Induction was first discovered way back in the 1830's by Michael Faraday. Faraday noticed that when he moved a permanent magnet in and out of a coil or a single loop of wire it induced an ElectroMotive Force or emf, in other words a Voltage, and therefore a current was produced.

So what Michael Faraday discovered was a way of producing an electrical current in a circuit by using only the force of a magnetic field and not batteries. This then lead to a very important law linking electricity with magnetism, Faraday's Law of Electromagnetic Induction. So how does this work?.

When the magnet shown below is moved "towards" the coil, the pointer or needle of the Galvanometer, which is basically a very sensitive centre zero'ed moving-coil ammeter, will deflect away from its centre position in one direction only. When the magnet stops moving and is held stationary with regards to the coil the needle of the galvanometer returns back to zero as there is no physical movement of the magnetic field.

Likewise, when the magnet is moved "away" from the coil in the other direction, the needle of the galvanometer deflects in the opposite direction with regards to the first indicating a change in polarity. Then by moving the magnet back and forth towards the coil the needle of the galvanometer will deflect left or right, positive or negative, relative to the directional motion of the magnet.

Electromagnetic induction is the process by which a current can be induced to flow due to a changing magnetic field.

In our article on the magnetic force we looked at the force experienced by moving charges in a magnetic field. The force on a current-carrying wire due to the electrons which move within it when a magnetic field is present is a classic example. This process also works in reverse. Either moving a wire through a magnetic field or (equivalently) changing the strength of the magnetic field over time can cause a current to flow.

HOW IS THIS DESCRIBED?

There are two key laws that describe electromagnetic induction:

1. Faraday's law, due to 19W"° century physicist Michael Faraday. This relates the rate of change of magnetic flux through a loop to the magnitude of the electro-motive force \mathcal{E}E induced in the loop. The relationship is \mathcal{E} = \frac{\mathrm{d}\Phi}{\mathrm{d}t}E= dt dÖ

The electromotive force or EMF refers to the potential difference across the unloaded loop (i.e. when the resistance in the circuit is high). In practice it is often sufficient to think of EMF as voltage since both voltage and EMF are measured using the same unit, the volt.

2. Lenz's law is a consequence of conservation of energy applied to electromagnetic induction. It was formulated by Heinrich Lenz in 1833. While Faraday's law tells us the magnitude of the EMF produced, Lenz's law tells us the direction that current will flow. It states that the direction is always such that it will oppose the change in flux which produced it. This means that any magnetic field produced by an induced current will be in the opposite direction to the change in the original field.

Lenz's law is typically incorporated into Faraday's law with a minus sign, the inclusion of which allows the same coordinate system to be used for both the flux and EMF. The result is sometimes called the Faraday-Lenz law,

\mathcal{E} = -\frac{\mathrm{d}\Phi}{\mathrm{d}t}E=" dt dÖ

In practice we often deal with magnetic induction in multiple coils of wire each of which contribute the same EMF. For this reason an additional term NNN representing the number of turns is often included, i.e.

\mathcal{E} = -N \frac{\mathrm{d}\Phi}{\mathrm{d}t}E="N dt dÖ

What is the connection between Faraday's law of induction and the magnetic force? While the full theoretical underpinning of Faraday's law is quite complex, a conceptual understanding of the direct connection to the magnetic force on a charged particle is relatively straightforward.

Consider an electron which is free to move within a wire. As shown in figure 1, the wire is placed in a vertical magnetic field and moved perpendicular to the magnetic field at constant velocity. Both ends of the wire are connected, forming a loop. This ensures that any work done in creating a current in the wire is dissipated as heat in the resistance of the wire.

A person pulls the wire with constant velocity through the magnetic field. As they do so, they have to apply a force. The constant magnetic field can't do work by itself (otherwise its strength would have to change), but it can change the direction of a force. In this case some of the force that the person applies is re-directed, causing an electromotive force on the electron which travels in the wire, establishing a current. Some of the work the person has done pulling the wire ultimately results in energy dissipated as heat within the resistance of the wire. Faraday's experiment : Induction from a magnet moving through a coil The key experiment which lead Michael Faraday to determine Faraday's law was quite simple. It can be quite easily replicated with little more than household materials. Faraday used a cardboard tube with insulated wire wrapped around it to form a coil. A voltmeter was connected across the coil and the induced EMF read as a magnet was passed through the coil.

THE OBSERVATIONS WERE AS FOLLOWS:

1. Magnet at rest in or near the coil: No voltage observed.
2. Magnet moving toward the coil: Some voltage measured, rising to a peak as the magnet nears the center of the coil.

3. Magnet passes through the middle of the coil: Measured voltage rapidly changes sign.
4. Magnet passes out and away from the coil: Voltage measured in the opposite direction to the earlier case of the magnet moving into the coil.

These observations are consistent with Faraday's law. Although the stationary magnet might produce a large magnetic field, no EMF can be induced because the flux through the coil is not changing. When the magnet moves closer to the coil the flux rapidly increases until the magnet is inside the coil. As it passes through the coil the magnetic flux through the coil begins to decrease. Consequently, the induced EMF is reversed.

Consider an electron which is free to move within a wire. As shown in figure 1, the wire is placed in a vertical magnetic field and moved perpendicular to the magnetic field at constant velocity. Both ends of the wire are connected, forming a loop. This ensures that any work done in creating a current in the wire is dissipated as heat in the resistance of the wire.

A person pulls the wire with constant velocity through the magnetic field. As they do so, they have to apply a force. The constant magnetic field can't do work by itself (otherwise its strength would have to change), but it can change the direction of a force. In this case some of the force that the person applies is re-directed, causing an electromotive force on the electron which travels in the wire, establishing a current. Some of the work the person has done pulling the wire ultimately results in energy dissipated as heat within the resistance of the wire.

Faraday's experiment : Induction from a magnet moving through a coil

The key experiment which lead Michael Faraday to determine Faraday's law was quite simple. It can be quite easily replicated with little more than household materials. Faraday used a cardboard tube with insulated wire wrapped around it to form a coil. A voltmeter was connected across the coil and the induced EMF read as a magnet was passed through the coil.

When Michael Faraday made his discovery of electromagnetic induction in 1831, he hypothesized that a changing magnetic field is necessary to induce a current in a nearby circuit. To test his hypothesis he made a coil by wrapping a paper cylinder with wire. He connected the coil to a galvanometer, and then moved a magnet back and forth inside the cylinder.

When you move the magnet back and forth, notice that the galvanometer needle moves, indicating that a current is induced in the coil. Notice also that the needle immediately returns to zero when the magnet is not moving. Faraday confirmed that a moving magnetic field is necessary in order for electromagnetic induction to occur.

In an entry in his notebook dated 17 October 1831, Faraday described how he made a cylindrical coil of copper wire wound round a paper tube. In those days one could not purchase insulated wire, and so the coil turns had to be separated from one another by means of string wound on the tube at the same time as the wire. Successive layers of turns were separated by strips of calico. When the coil was finished its ends were connected to a sensitive galvanometer. Faraday then plunged a magnet into the coil and noticed that the galvanometer needle gave a momentary deflection, showing that a current had been induced in the coil. On removal of the magnet from the coil the galvanometer gave another deflection, this time in the opposite direction (Fig. 43.1). This effect is called electromagnetic induction. No current was induced imply by allowing the magnet to remain at rest inside or outside the coil.

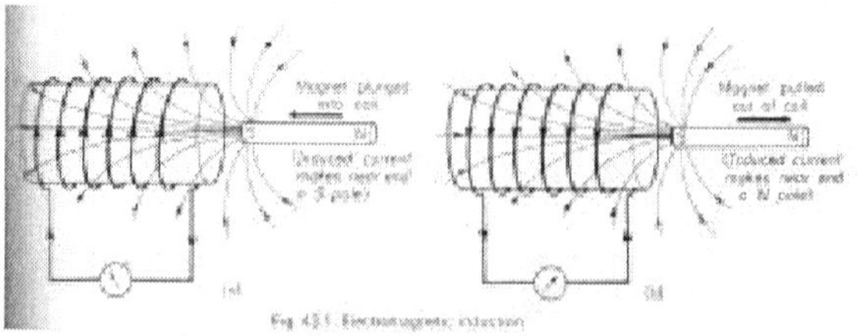

Faraday's experiments on electromagnetic induction

After much thought Faraday came to the conclusion that current was induced in the circuit as a result of the wire being cut hy magnetic/lux when the magnet moved. Another way of describing this phenomenon is to say that an electromotive force is – duced whenever there is a change in the magnetic flux linked with the coil. Continuing his investigations with other coils and magnets, Faraday found that e strength of the induced current depends on:

(I) the number of turns in the coil;

(2) the strength of the magnet;

(3) the speed with which the magnet is plunged into the coil.

In another experiment Faraday wound a coil on a bar of soft iron and connected e ends of the coil to a galvanometer (Fig. 43.2). The opposite poles of a pair of

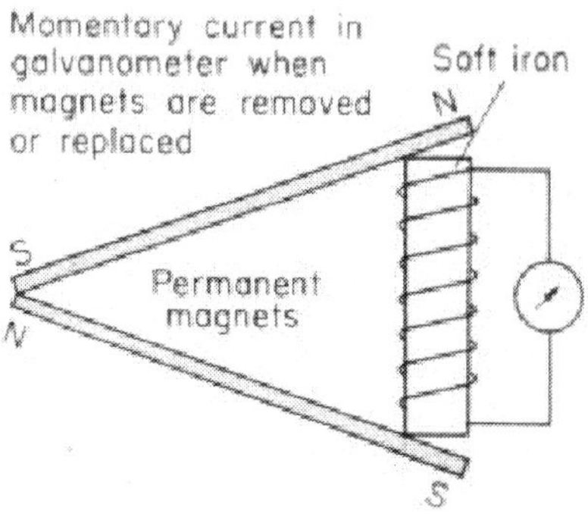

Faraday's View

were then brought into contact with the ends of the iron bar. This produced rge concentration of magnetic flux through the coil, and the galvanometer gave a , showing that a momentary current had been induced. On removing the maga current was induced in the coil in the opposite direction. The above experiment will also work without iron inside the coil, but in this ase the induced current is much weaker. The effect of the iron is to increase the magnetic flux through the coil, and consequently the induced electromotive force is greater. A New Look at Electric Potential Induced electric fields are produced not by static charges but by a changing magnetic flux. Although electric fields produced in either way exert forces on charged particles. there is an important difference between them. The simplest evidence of this difference is that the field lines of induced electric fields form closed loops, as in charges. In a more formal sense, we can state the difference between electric fields Equation 31-27 gives the magnitude of the electric field at any point for which r S R (that is, within the magnetic field). Substituting given values yields, for the magnitude of Eat r = 5.2 em,

E = (5.2 X 10-2 m) (0.13 TIs)

- While studying magnetism we learned that electricty and magnetism are interrelated and infact moving charges or electric currents produces magnetic field that can deflect magnetic compass needle
- A question arises,can moving magnetic field produce electricty. Answer is yes and it was first showed by a british scientist Michael faraday in 1931 who after performing various experimentsfound that moving magnetic field can give rise to the EMF
- Independently the effect was discovered by Joseph Henry in USA at about the same time
- In this chapter we will discuss about the electric and magnetic field changing with time
- More precisely we will consider the phenomenon related to time changing current or time changing magnetic fields
- faraday in 1831 first discovered that whenever the number it of magnetic lines of forces in a circuit changes, a emf is produced in the circuit and is known as induced emf and this phenomenon is known as Electro Magnetic Induction
- If the circuit is closed one then a current flows through it which is known an induced current
- This induced emf and current lasts only for the time while magnetic flux is chnaging
- We now illustrate two examples of the sort that faraday and henry performed

  2= 0.0034 VIm = 3.4 mV/m

  Experiments of Faraday and Henry

OCTOBER 5, 20170 COMMENTS

In this section we will learn about the experiments carried out by Faraday and Henry that are used to understand the phenomenon of electromagnetic induction and its properties.

EXPERIMENT 1:

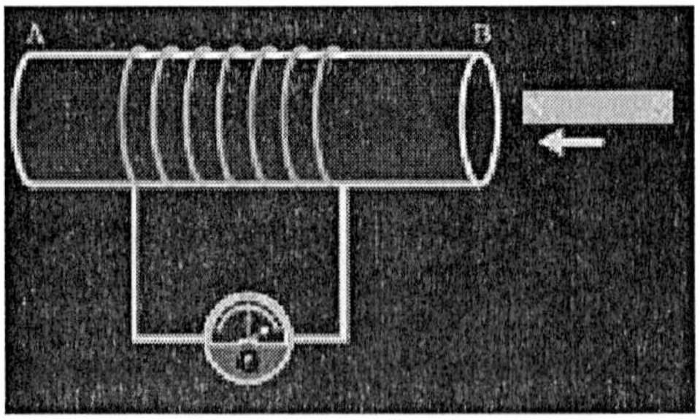

In this experiment, Faraday connected a coil to a galvanometer, as shown in the figure above. A bar magnet was pushed towards the coil, such that the north-pole is pointing towards the coil. As the bar magnet is shifted, the pointer in the galvanometer gets deflected, thus indicating the presence of current in the coil under consideration. It is observed that when the bar magnet is stationary, the pointer shows no deflection and the motion lasts only till the magnet is in motion. Here, the direction of deflection of the pointer depends upon the direction of motion of the bar magnet. Also, when the south-pole of the bar magnet is moved towards or away from the coil, the deflections in the galvanometer are opposite to that observed with the north-pole for similar movements. Apart from this, the deflection of the pointer is larger or smaller depending upon the speed with which it is pulled towards or away from the coil. The same effect is observed when instead of the bar magnet, the coil is moved and the magnet is held stationary. This shows that only the relative motion between the magnet and the coil is responsible for the generation of current in the coil.

EXPERIMENT 2:

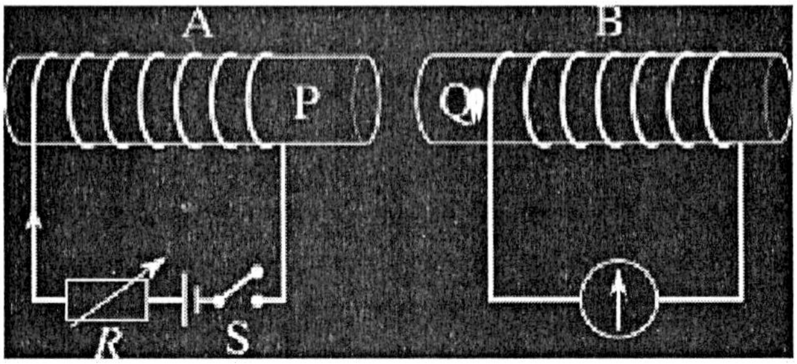

In the second experiment, Faraday replaced the bar magnet by a second current carrying coil that was connected to a battery. Here, the current in the coil due to the connected battery produced a steady magnetic field, which made the system analogous to the previous one. As we move the second coil towards the primary coil, the pointer in the galvanometer undergoes deflection, which indicates the presence of the electric current in the first coil. Similar to the above case, here too, the direction of deflection of the pointer depends upon the direction of motion of the secondary coil towards or away from the primary coil. Also, the magnitude of deflection depends upon the speed with which the coil is moved. All these results show that the system in the second case is analogous to the system in the first experiment.

EXPERIMENT 3:

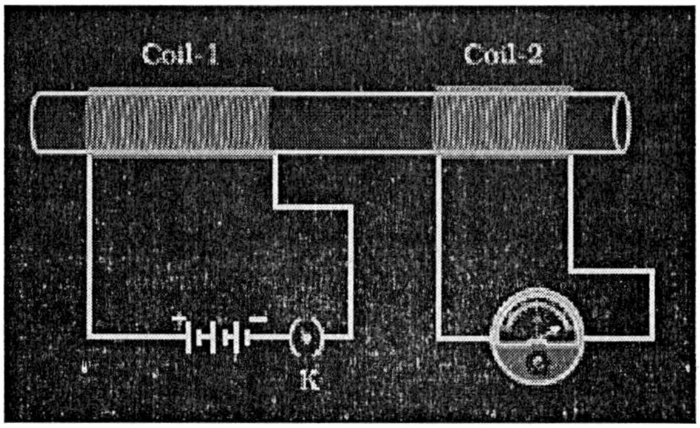

From the above two experiments, it was concluded by Faraday that the relative motion between the magnet and the coil resulted in the generation of current in the primary coil. But another experiment conducted by Faraday proved that the relative motion between the coils was not really necessary for the current in the primary to be generated. In this experiment he placed two stationary coils and connected one of them to the galvanometer and the other to a battery, through a push button. As the button was pressed, the galvanometer in the other coil showed a deflection, indicating the presence of current in that coil. Also, the deflection in the pointer was temporary and if pressed continuously, the pointer showed no deflection and when the key was released, the deflection occurred in the opposite direction.

Stay tuned with byju's to learn more about the experiments by Faraday and Henry, relative motion and much more.

5

Faradays' Law and it's Relativity

THE FARADAY INDUCTION LAW IN RELATIVITY THEORY

We analyze the transformation properties of Faraday's law in an empty space and its relationship with Maxwell's equations. In our analysis we express the Faraday's law via the fourpotential of electromagnetic field and the field of four-velocity, defined on a circuit under its deforming motion. The obtained equations show one more facet of the physical meaning of electromagnetic potentials, where the motional and transformer parts of the flux rule are incorporated into a common phenomenon, reflecting the dependence of four-potential on spatial and time coordinates, correspondingly. It has been explicitly shown that for filamentary closed deforming circuit the flux rule is Lorentz-invariant. At the same time, analyzing a transformation of e.m.f., we revealed a controversy: due to causal requirements, the e.m.f. should be the value of fixed sign, whereas the Lorentz invariance of flux rule admits the cases, where the e.m.f. might change its sign for different inertial observers. Possible resolutions of this controversy are discussed.

It is known that Faraday's law involves two different phenomena in a single equation: law of induction ("transformer" part of induction), and Lorentz force law ("motional" part of induction, when the circuit moves). Numerous papers and books were devoted to Faraday's law. However, its transformation properties were analyzed up to date only in a few papers (see, e.g., refs. [2-6]). The case of time-variable contour $\tilde{A}=\tilde{A}(t)$ happens to be especially difficult for such an analysis, and some ambiguities continue to exist up to now. In particular, one can mention the questions as follows: 1. Is the empirically discovered Faraday's law.

What's the right way to divide these sections up?

MOTIONAL EMF (ELECTROMOTIVE FORCE)

0! = B–!v

ELECTROMAGNETIC INDUCTION AND EMF

$$\mathcal{E} = -\frac{\Delta\Phi_B}{\Delta t}$$

$$\mathcal{E} = -\frac{d\Phi_B}{dt}$$

A more advanced form written as an bunch of integrals

$$\oint \mathbf{E} \cdot d\mathbf{s} = -\frac{\partial\Phi_B}{\partial t}$$

A more advanced form written as a bunch of derivatives

$$\nabla \times \mathbf{E} = -\frac{\partial \mathbf{B}}{\partial t}$$

Lenz' law: Induced current flows in the direction that opposes the change that created it.

APPLICATIONS

During top level geomagnetic storms (G5 out of 5) pipeline currents can reach hundreds of amps. On average these occur 4 times every 11 year solar cycle and last about 4 days each.

A GENERATOR IS A BACKWARD MOTOR.

A dynamic microphone is a backward speaker.

FARADAY QUOTE

Entry from Faraday's lab notebook dated 29 August 1931 as quoted in The Life and Letters of Faraday by Bence Jones. Philadelphia: Lippicott (1870): Vol. II, Page 2.

I have had an iron ring made (soft iron), iron round and ^!ths of an inch thick, and ring six inches in external diameter. Wound many coils of copper round, one half of the coils being separated by twine and calico; there were three lengths of wire, each about twenty-four feet long, and they could be connected as one length, or used as separate lengths. By trials with a trough each was insulated from the other. Will call this side of the ring A. On the other side, but separated by an interval, was wound wire in two pieces, together amounting to about sixty feet in length, the direction being as with the former coils. This side call B. Charged a battery of ten pairs of plates four inches square. Made the coil on B side one coil, and connected its extremities by a copper wire passing to a distance, and just over a magnetic needle (three feet from wire ring), then connected the ends of one of the pieces on A side with battery: immediately a sensible effect on needle. It oscillated and settled at last in original position. On breaking connection of A side with battery, again a disturbance of the needle.

MORE FARADAY QUOTES

Philosophical Transactions, 1832, 122: 125-62 V. Experimental Researches in Electricity. By Michael Faraday, F.R.S., M.R.I., Corr. Mem. Royal Acad. of Sciences of Paris, Petersburgh, &c. &c. Read November 24, 1831.

[mutual induction] 10. Two hundred and three feet of copper wire in one length were passed round a large block of wood; other two hundred and three feet of similar wire were interposed as a spiral between the turns of the first, and metallic contact everywhere prevented by twine. One of these helices was connected with a galvanometer, and the other with a battery of one hundred pairs of plates four inches square, with double coppers, and well charged. When the contact was made, there was a sudden and very slight effect at the galvanometer, and there was also a similar slight effect when the contact with the battery was broken. But whilst the voltaic current was continuing to pass through the one helix, no galvanometrical appearances of any effect like induction upon the other helix could be perceived, although the active power of the battery was proved to be great, by its heating the whole of its own helix, and by the brilliancy of the discharge when made through charcoal.

[motional emf] 18. In the preceding experiments the wires were placed near to each other, and the contact of the inducing one with the battery made when the inductive effect was required; but as some particular action might be supposed to he exerted at the moments of making and breaking contact, the induction was produced in another way. Several feet of copper wire were stretched in wide zigzag forms, representing the letter W, on one surface of a broad board; a second wire was stretched in precisely similar forms on a second board, so that when brought near the first, the wires should everywhere touch, except that a sheet of thick paper was interposed. One of these wires was connected with the galvanometer, and the other with a voltaic battery. The first wire was then moved towards the second, and as it approached, the needle was deflected. Being then removed, the needle was deflected in the opposite direction. By first making the wires approach and then recede, simultaneously with the vibrations of the needle, the latter soon became very extensive; but when the wires ceased to move from or towards each other, the galvanometer needle soon came to its usual position.

[iron core electromagnet] 34. Another arrangement was then used connecting the former experiments on volta-electric induction with the present.

A combination of helices like that already described (6.) was constructed upon a hollow cylinder of pasteboard: there were eight lengths of copper wire, containing altogether 220 feet; four of these helices were connected end to end, and then with the galvanometer (7.); the other intervening four were also connected, end to end, and the battery of one hundred pairs discharged through them. In this form the effect on the galvanometer was hardly sensible (11.), but magnets could be made by the induced current (13.). But when a soft iron cylinder seven eighths of an inch thick, and twelve inches long, was introduced into the pasteboard tube; surrounded by the helices, then the induced current affected the galvanometer powerfully, and with all the phenomena just described (30.). It possessed also the power of making magnets with more energy, apparently, than when no iron cylinder was present.

When the iron cylinder was replaced by an equal cylinder of copper, no effect beyond that of the helices alone was produced. The iron cylinder arrangement was not so powerful as the ring arrangement already described (27.).

[motional emf] 36. Similar effects were then produced by ordinary magnets: thus the hollow helix just described (34.) had all its elementary helices connected with the galvanometer by two copper wires, each five feet in length; the soft iron cylinder was introduced into its axis; a couple of bar magnet; each twenty-four inches long, were arranged with their opposite poles at one end in contact, so as to resemble a horse-shoe magnet, and then contact made between the other poles and the ends of the iron cylinder, so as to convert it for the time into a magnet (fig.2.): by breaking the magnetic contacts, or reversing them, the magnetism of the iron cylinder could be destroyed or reversed at pleasure.

Upon making magnetic contact, the needle was deflected; continuing the contact, the needle became indifferent, and resumed its first position; on breaking the contact, it was again deflected, but in the opposite direction to the first effect, and then it again became indifferent When the magnetic contacts were reversed, the deflections were reversed.

When the magnetic contact was made, the deflection was such as to indicate an induced current of electricity in the opposite direction to that fitted to form a magnet having the same polarity as that really produced by contact with the bar magnets. Thus when the marked and unmarked poles were placed as in fig.3, the current in the helix was in the direction represented, P being supposed to be the end of the wire going to the positive pole of the battery, or that end towards which the zinc plates face, and N the negative wire. Such a current would have converted the cylinder into a magnet of the opposite kind to that formed by contact with the poles A and B; and such a current moves in the opposite direction to the currents whet in M. Ampere's beautiful theory are considered as constituting a magnet in the position figured.

But as it might be supposed that in all the preceding experiment of this section it was by some peculiar effect taking place during the formation of the magnet, and not by its mere virtual approximation, that the momentary induced current was excited, the following experiment was made. All the similar ends of the compound hollow helix (34.) were bound together by copper wire, forming two general terminations, and these were connected with the galvanometer. The soft iron cylinder (34.) was removed, and a cylindrical magnet, three quarters of an inch in diameter and eight inches and a half in length, used instead. One end of this magnet was introduced into the axis of the helix (fig.4.), and then, the galvanometer-needle being stationary, the magnet was suddenly thrust in; immediately the needle was deflected in the same direction as if the magnet had been formed by either of the two preceding processes (34. 36.). Being left in, the needle resumed its first position, and then the magnet being withdrawn the needle was deflected in the opposite direction. These effects were not great; but by introducing and withdrawing the magnet, so that the impulse each time should be added to those previously communicated to the needle, the latter could be made to vibrate through an arc of 180° or more.

In this experiment the magnet must not be passed entirely through the helix, for then a second action occurs. When the magnet is introduced, the needle at the galvanometer is deflected in a certain direction; but being in, whether it be pushed quite through or withdrawn, the needle is deflected in a direction the reverse of that previously produced. When the magnet is passed in and through at one continuous motion, the needle moves one way, is then suddenly stopped, and finally moves the other way.

The Faraday paradox or Faraday's paradox is any experiment in which Michael Faraday's law of electromagnetic induction appears to predict an incorrect result. The paradoxes fall into two classes:

- Faraday's law appears to predict that there will be zero EMF but there is a non-zero EMF.
- Faraday's law appears to predict that there will be a non-zero EMF but there is zero EMF.

Faraday deduced his law of induction in 1831, after inventing the first electromagnetic generator or dynamo, but was never satisfied with his own explanation of the paradox.

Faraday's law compared to the Maxwell–Faraday equation

Faraday's law (also known as the Faraday–Lenz law) states that the electromotive force (EMF) is given by the total derivative of the magnetic flux with respect to time t:{\displaystyle {\mathcal {E}}=-{d\Phi _{B} \over dt},\ }

where {\displaystyle {\mathcal {E}}}is the EMF and Ö$_B$ is the magnetic flux. The direction of the electromotive force is given by Lenz's law. An often overlooked fact is that Faraday's law is based on the total derivative, not the partial derivative, of the magnetic flux. This means that an EMF may be generated even if total flux through the surface is constant. To overcome this issue, special techniques may be used. See below for the section on Use of special techniques with Faraday's law. However, the most common interpretation of Faraday's law is that:

The induced electromotive force in any closed circuit is equal to the negative of the time rate of change of the magnetic flux enclosed by the circuit.

The Maxwell–Faraday equation is a generalisation of Faraday's law that states that a time-varying magnetic field is always accompanied by a spatially-varying, non-conservativeelectric field, and vice versa.

(in SI units) where {\displaystyle \partial } is the partial derivative operator, {\displaystyle \nabla \times }is the curl operator and again E(r, t) is the electric field and B(r, t) is the magnetic field. These fields can generally be functions of position r and time t.

The Maxwell–Faraday equation is one of the four Maxwell's equations, and therefore plays a fundamental role in the theory of classical electromagnetism. It can also be written in an integral form by the Kelvin–Stokes theorem.

Paradoxes in which Faraday's law of induction seems to predict zero EMF but actually predicts non-zero EMF

These paradoxes are generally resolved by the fact that an EMF may be created by a changing flux in a circuit as explained in Faraday's law or by the movement of a conductor in a magnetic field. This is explained by Feynman as noted below. See also A. Sommerfeld, Vol III Electrodynamics Academic Press, page 362.

THE EQUIPMENT

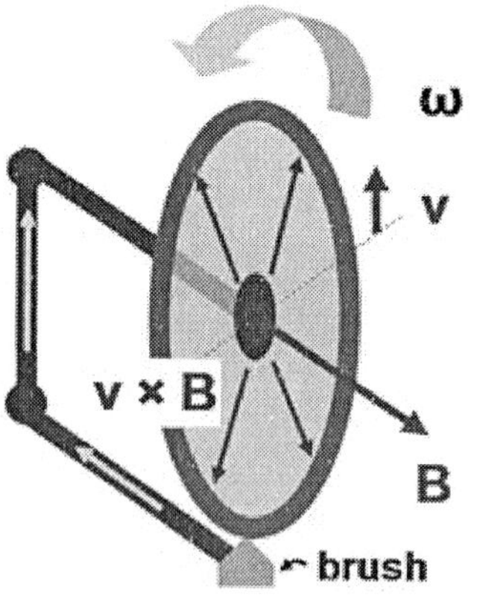

Figure 1: Faraday's disc electric generator. The disc rotates with angular rate ù, sweeping the conducting disc circularly in the static magnetic field B due to a permanent magnet. The magnetic Lorentz force v × B drives the current

radially across the conducting disc to the conducting rim, and from there the circuit path completes through the lower brush and the axle supporting the disc. Thus, current is generated from mechanical motion.

The experiment requires a few simple components (see Figure 1): a cylindrical magnet, a conducting disc with a conducting rim, a conducting axle, some wiring, and a galvanometer. The disc and the magnet are fitted a short distance apart on the axle, on which they are free to rotate about their own axes of symmetry. An electrical circuit is formed by connecting sliding contacts: one to the axle of the disc, the other to its rim. A galvanometer can be inserted in the circuit to measure the current.

THE PROCEDURE

The experiment proceeds in three steps:

1. The magnet is held to prevent it from rotating, while the disc is spun on its axis. The result is that the galvanometer registers a direct current. The apparatus therefore acts as a generator, variously called the Faraday generator, the Faraday disc, or the homopolar (or unipolar) generator.
2. The disc is held stationary while the magnet is spun on its axis. The result is that the galvanometer registers no current.
3. The disc and magnet are spun together. The galvanometer registers a current, as it did in step 1.

WHY IS THIS PARADOXICAL?

The experiment is described by some as a "paradox" as it seems, at first sight, to violate Faraday's law of electromagnetic induction, because the flux through the disc appears to be the same no matter what is rotating. Hence, the EMF is predicted to be zero in all three cases of rotation. The discussion below shows this viewpoint stems from an incorrect choice of surface over which to calculate the flux.

The paradox appears a bit different from the lines of flux viewpoint: in Faraday's model of electromagnetic induction, a magnetic field consisted of imaginary lines of magnetic flux, similar to the lines that appear when iron filings are sprinkled on paper and held near a magnet. The EMF is proposed to be proportional to the rate of cutting lines of flux. If the lines of flux are imagined to originate in the magnet, then they would be stationary in the frame of the magnet, and rotating the disc relative to the magnet, whether by rotating the magnet or the disc, should produce an EMF, but rotating both of them together should not.

FARADAY'S EXPLANATION

In Faraday's model of electromagnetic induction, a circuit received an induced current when it cut lines of magnetic flux. According to this model, the Faraday disc should have worked when either the disc or the magnet was rotated, but not both. Faraday attempted to explain the disagreement with observation by assuming that the magnet's field, complete with its lines of flux, remained stationary as the magnet rotated (a completely accurate picture, but maybe not intuitive in the lines-of-flux model). In other words, the lines of flux have their own frame of reference. As we shall see in the next section, modern physics (since the discovery of the electron) does not need the lines-of-flux picture and dispels the paradox.

MODERN EXPLANATIONS

TAKING THE RETURN PATH INTO ACCOUNT

In step 2, since there is no current observed, one might conclude that the magnetic field did not rotate with the rotating magnet. (Whether it does or does not effectively or relatively, the Lorentz force is zero since v is zero relative to the laboratory frame. So there is no current measuring from laboratory frame.) The use of the Lorentz equation to explain this paradox has led to a debate in the literature as to whether or not a magnetic field rotates with a magnet. Since the force on charges expressed by the Lorentz equation depends upon the relative motion of the magnetic field (i.e. the laboratory frame) to the conductor where the EMF is located it was speculated that in the case when the magnet rotates with the disk but a voltage still develops, the magnetic field (i.e. the laboratory frame) must therefore not rotate with the magnetic material (of course since it is the laboratory frame), while the effective definition of magnetic field frame or the "effective/relative rotation of the field" turns with no relative motion with respect to the conductive disk.

Careful thought showed that, if the magnetic field was assumed to rotate with the magnet and the magnet rotated with the disk, a current should still be produced, not by EMF in the disk (there is no relative motion between the disk and the magnet) but in the external circuit linking the brushes, which is in fact in relative motion with respect to the rotating magnet. (The brushes are in the laboratory frame.)

This mechanism agrees with the observations involving return paths: an EMF is generated whenever the disc moves relative to the return path, regardless of the rotation of the magnet. In fact it was shown that so long as a current loop is used to measure induced EMFs from the motion of the disk and magnet it is not possible to tell if the magnetic field does or does not rotate with the magnet. (This depends on the definition, the motion of a field can be only defined effectively/relatively. If you hold the view that the field flux is a physical entity, it does rotate or depends on how it is generated. But this does not alter what is used in the Lorentz formula, especially the v, the velocity of the charge carrier relative to the frame where measurement takes place and field strength varies according to relativity at any spacetime point.)

Several experiments have been proposed using electrostatic measurements or electron beams to resolve the issue, but apparently none have been successfully performed to date.

When the magnet is rotating, but flux lines are stationary, and the conductor is stationary

Consider the special case where the cylindrical conducting disk is stationary but the cylindrical magnetic disk is rotating. In such a situation, the mean velocity v of charges in the conducting disk is initially zero, and therefore the magnetic force F = qv × B is 0, where v is the mean velocity of a charge q of the circuit relative to the frame where measurements are taken, and q is the charge on an electron.

Note that v does not represent the velocity that magnetic fields lines travel through the conductor. The magnetic field pattern observed depends on the frame of reference. They have no velocity on their own. To illustrate, imagine that one were to take a magnet and flip it 180 degrees. The result of assuming that these field lines actually have velocity of their own would be that magnetic field lines far away would swing to the other side, potentially faster than the speed of light, however, that is not what happens.

Instead, what happens is that rotating a magnet causes subatomic particles in the magnet to acquire a change in velocity. However, in the rotated magnet, the velocities of electrons would vary ahead or behind those of the other particles, due to their lower mass compared to nuclei. There is length contraction of fields propagating from moving charges, and the length contraction of the electric fields E of the electrons would be greater or lesser than the contraction of the E fields of the positive nuclei depending on whether the rotation of the magnet was aligned to, or opposed to, the electron spins which give rise to the magnetism.

In the case of an axially-symmetric magnet spinning at constant velocity, the distribution of the magnetic field intensity B of the magnet is constant with time, even after accounting for the relativistic corrections to B, and therefore by the Maxwell–Faraday equation the electric field induced by the rotation of the magnet is curl-free, which is caused purely by length contraction of the electric fields of its constituting subatomic particles. This means that in this particular example with a rotating magnetic disk and a stationary conducting disk, the induced electric field cannot be described by the Maxwell–Faraday equation, which describes the curl of the electric field induced by a changing magnetic flux density.

Therefore, in this view, magnetic fields do not rotate with their magnetic source, and they exist independently of them. Nevertheless, the Lorentz forces generated by the rotation of the magnet are as if these lines rotated in unison with it, but this is in reality due to the effect of relativity on electric fields.

When the magnet and the flux lines are stationary and the conductor is rotating

After the discovery of the electron and the forces that affect it, a microscopic resolution of the paradox became possible. See Figure 1. The metal portions of the apparatus are conducting, and confine a current due to electronic motion to within the metal boundaries. All electrons that move in a magnetic field experience a Lorentz force of F = qv × B, where v is the velocity of the electrons relative to the frame where measurements are taken, and q is the charge on an electron. Remember, there is no such frame as "frame of the electromagnetic field". A frame is set on a specific spacetime point, not an extending field or a flux line as a mathematical object. It is a different issue if you consider flux as a physical entity (see Magnetic flux quantum), or consider the effective/relative definition of motion/rotation of a field (see below). This note helps resolve the paradox.

The Lorentz force is perpendicular to both the velocity of the electrons, which is in the plane of the disc, and to the magnetic field, which is normal (surface normal) to the disc. An electron at rest in the frame of the disc moves circularly with the disc relative to the B-field (i.e. the rotational axis or the laboratory frame, remember the note above), and so experiences a radial Lorentz force. In Figure 1 this force (on a positive charge, not an electron) is outward toward the rim according to the right-hand rule. Of course, this radial force, which is the cause of the current, creates a radial component of electron velocity, generating in turn its own Lorentz force component that opposes the circular motion of the electrons, tending to slow the disc's rotation, but the electrons retain a component of circular motion that continues to drive the current via the radial Lorentz force.

In choosing the surface Ó(t), the restrictions are that (i) it has to be bounded by a closed curve around which the EMF is to be found, and (ii) it has to capture the relative motion of all moving parts of the circuit. It is emphatically not required that the bounding curve corresponds to a physical line of flow of the current. On the other hand, induction is all about relative motion, and the path emphatically must capture any relative motion. In a case like Figure 1 where a portion of the current path is distributed over a region in space, the EMF driving the current can be found using a variety of paths. Figure 2 shows two possibilities. All paths include the obvious return loop, but in the disc two paths are shown: one is a geometrically simple path, the other a tortuous one. We are free to choose whatever path we like, but a portion of any acceptable path is fixed in the disc itself and turns with the disc. The flux is calculated though the entire path, return loop plus disc segment, and its rate-of change found.

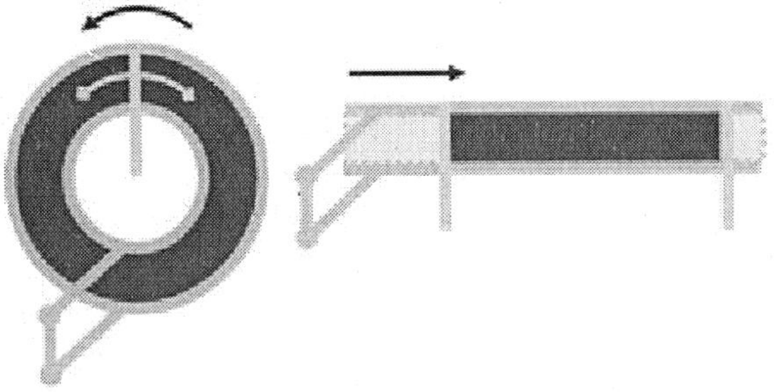

Figure 3: Mapping of the Faraday disc into a sliding conducting rectangle example. The disc is viewed as an annulus; it is cut along a radius and bent open to become a rectangle.

In this example, all these paths lead to the same rate of change of flux, and hence the same EMF. To provide some intuition about this path independence, in Figure 3 the Faraday disc is unwrapped onto a strip, making it resemble a sliding rectangle problem. In the sliding rectangle case, it becomes obvious that the pattern of current flow inside the rectangle is time-independent and therefore irrelevant to the rate of change of flux linking the circuit. There is no need to consider exactly how the current traverses the rectangle (or the disc). Any choice of path connecting the top and bottom of the rectangle (axle-to-brush in the disc) and moving with the rectangle (rotating with the disc) sweeps out the same rate-of-change of flux, and predicts the same EMF. For the disc, this rate-of-change of flux estimation is the same as that done above based upon rotation of the disc past a line joining the brush to the axle.

CONFIGURATION WITH A RETURN PATH

Whether the magnet is "moving" is irrelevant in this analysis, due the flux induced in the return path. The crucial relative motion is that of the disk and the return path, not of the disk and the magnet. This becomes clearer if a modified Faraday disk is used in which the return path is not a wire but another disk. That is, mount two conducting disks just next to each other on the same axle and let them have sliding electrical contact at the center and at the circumference. The current will be proportional to the relative rotation of the two disks and independent of any rotation of the magnet.

CONFIGURATION WITHOUT A RETURN PATH

A Faraday disk can also be operated with neither a galvanometer nor a return path. When the disk spins, the electrons collect along the rim and leave a deficit near the axis (or the other way around). It is possible in principle to measure the distribution of charge, for example, through the electromotive force generated between the rim and the axle (though not necessarily easy). This charge separation will be proportional to the relative rotational velocity between the disk and the magnet.

Paradoxes in which Faraday's law of induction seems to predict non-zero EMF but actually predicts zero EMF

These paradoxes are generally resolved by determining that the apparent motion of the circuit is actually deconstruction of the circuit followed by reconstruction of the circuit on a different path.

AN ADDITIONAL RULE

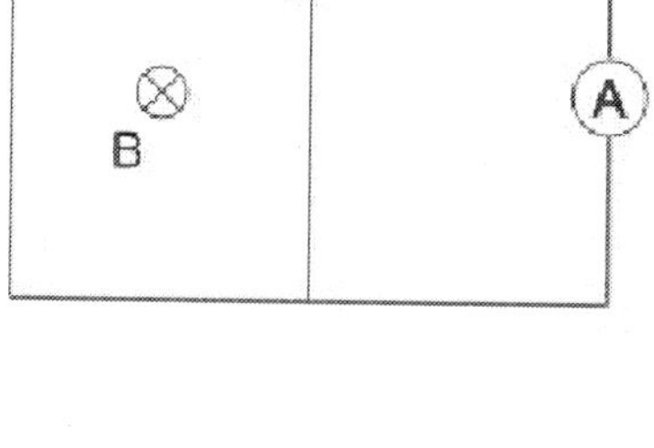

In the case when the disk alone spins there is no change in flux through the circuit, however, there is an electromotive force induced contrary to Faraday's law. We can also show an example when there is a change in flux, but no induced voltage. Figure 5 (near right) shows the setup used in Tilley's experiment. It is a circuit with two loops or meshes. There is a galvanometer connected in the right-hand loop, a magnet in the center of the left-hand loop, a switch in the left-hand loop, and a switch between the loops. We start with the switch on the left open and that on the right closed. When the switch on the left is closed and the switch on the right is open there is no change in the field of the magnet, but there is a change in the area of the galvanometer circuit. This means that there is a change in flux. However the galvanometer did not deflect meaning there was no induced voltage, and Faraday's law does not work in this case. According to A. G. Kelly this suggests that an induced voltage in Faraday's experiment is due to the "cutting" of the circuit by the flux lines, and not by "flux linking" or the actual change in flux. This follows from the Tilley experiment because there is no movement of the lines of force across the circuit and therefore no current induced although there is a change in flux through the circuit. Nussbaum suggests that for Faraday's law to be valid, work must be done in producing the change in flux.To understand this idea, we will step through the argument given by Nussbaum. We start by calculating the force between two current carrying wires.

The galvanometer usually only measures the first term in the EMF which contributes the current in circuit, although sometimes it can measure the incorporation of the second term such as when the second term contributes part of the current which the galvanometer measures as motional EMF, e.g. in the Faraday's disk experiment. In the situation above, the first term is zero and only the first term leads a current that the galvanometer measures, so there is no induced voltage. However, Faraday's Law still holds since the apparent change of the magnetic flux goes to the second term in the above generalization of EMF. But it is not measured by the galvanometer. It is the local velocity of a point on the circuit, not a charge carrier. After all, both/all these situations are consistent with the concern of relativity and microstructure of matter, and/or the completeness of Maxwell equation and Lorentz formula, or the combination of them, Hamiltonian mechanics.

MOVING MAGNET AND CONDUCTOR PROBLEM

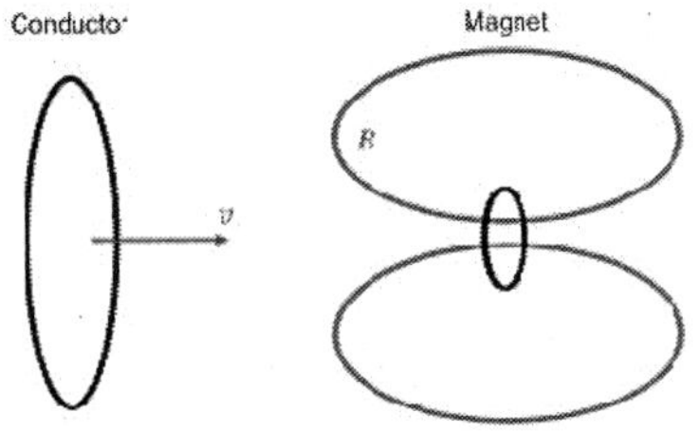

Conductor moving in a magnetic field.

The moving magnet and conductor problem is a famous thought experiment, originating in the 19th century, concerning the intersection of classical electromagnetism and special relativity. In it, the current in a conductor moving with constant velocity, v, with respect to a magnet is calculated in the frame of reference of the magnet and in the frame of reference of the conductor. The observable quantity in the experiment, the current, is the same in either case, in accordance with the basic principle of relativity, which states: "Only relative motion is observable; there is no absolute standard of rest". However, according to Maxwell's equations, the charges in the conductor experience a magnetic force in the frame of the magnet and an electric force in the frame of the conductor. The same phenomenon would seem to have two different descriptions depending on the frame of reference of the observer.

This problem, along with the Fizeau experiment, the aberration of light, and more indirectly the negative aether drift tests such as the Michelson–Morley experiment, formed the basis of Einstein's development of the theory of relativity.

Einstein's 1905 paper that introduced the world to relativity opens with a description of the magnet/conductor problem.

It is known that Maxwell's electrodynamics – as usually understood at the present time – when applied to moving bodies, leads to asymmetries which do not appear to be inherent in the phenomena. Take, for example, the reciprocal electrodynamic action of a magnet and a conductor. The observable phenomenon here depends only on the relative motion of the conductor and the magnet, whereas the customary view draws a sharp distinction between the two cases in which either the one or the other of these bodies is in motion. For if the magnet is in motion and the conductor at rest, there arises in the neighborhood of the magnet an electric field with a certain definite energy, producing a current at the places where parts of the conductor are situated. But if the magnet is stationary and the conductor in motion, no electric field arises in the neighborhood of the magnet. In the conductor, however, we find an electromotive force, to which in itself there is no corresponding energy, but which gives rise – assuming equality of relative motion in the two cases discussed – to electric currents of the same path and intensity as those produced by the electric forces in the former case.

A. EINSTEIN, ON THE ELECTRODYNAMICS OF MOVING BODIES (1905)

An overriding requirement on the descriptions in different frameworks is that they be consistent. Consistency is an issue because Newtonian mechanics predicts one transformation (so-called Galilean invariance) for the forces that drive the charges and cause the current, while electrodynamics as expressed by Maxwell's equations predicts that the fields that give rise to these forces transform differently (according to Lorentz invariance). Observations of the aberration of light, culminating in the Michelson–Morley experiment, established the validity of Lorentz invariance, and the development of special relativity resolved the resulting disagreement with Newtonian mechanics. Special relativity revised the transformation of forces in moving reference frames to be consistent with Lorentz invariance. The details of these transformations are discussed below.

In addition to consistency, it would be nice to consolidate the descriptions so they appear to be frame-independent. A clue to a framework-independent description is the observation that magnetic fields in one reference frame become electric fields in another frame. Likewise, the solenoidal portion of electric fields (the portion that is not originated by electric charges) becomes a magnetic field in another frame: that is, the solenoidal electric fields and magnetic fields are aspects of the same thing. That means the paradox of different descriptions may be only semantic. A description that uses scalar and vector potentials ö and A instead of B and E avoids the semantical trap. A Lorentz-invariant four vector A = (ö / c, A) replaces E and B and provides a frame-independent description (albeit less visceral than the E– B– description). An alternative unification of descriptions is to think of the physical entity as the electromagnetic field tensor, as described later on. This tensor contains both E and B fields as components, and has the same form in all frames of reference.

BACKGROUND

Electromagnetic fields are not directly observable. The existence of classical electromagnetic fields can be inferred from the motion of charged particles, whose trajectories are observable. Electromagnetic fields do explain the observed motions of classical charged particles.

A strong requirement in physics is that all observers of the motion of a particle agree on the trajectory of the particle. For instance, if one observer notes that a particle collides with the center of a bullseye, then all observers must reach the same conclusion. This requirement places constraints on the nature of electromagnetic fields and on their transformation from one reference frame to another. It also places constraints on the manner in which fields affect the acceleration and, hence, the trajectories of charged particles.

Perhaps the simplest example, and one that Einstein referenced in his 1905 paper introducing special relativity, is the problem of a conductor moving in the field of a magnet. In the frame of the magnet, a conductor experiences a magnetic force. In the frame of a conductor moving relative to the magnet, the conductor experiences a force due to an electric field. The magnetic field in the magnet frame and the electric field in the conductor frame must generate consistent results in the conductor. At the time of Einstein in 1905, the field equations as represented by Maxwell's equations were properly consistent. Newton's law of motion, however, had to be modified to provide consistent particle trajectories.

TRANSFORMATION OF FIELDS, ASSUMING GALILEAN TRANSFORMATIONS

Assuming that the magnet frame and the conductor frame are related by a Galilean transformation, it is straightforward to compute the fields and forces in both frames. This will demonstrate that the induced current is indeed the same in both frames. As a byproduct, this argument will also yield a general formula for the electric and magnetic fields in one frame in terms of the fields in another frame.

In reality, the frames are not related by a Galilean transformation, but by a Lorentz transformation. Nevertheless, it will be a Galilean transformation to a very good approximation, at velocities much less than the speed of light.

Unprimed quantities correspond to the rest frame of the magnet, while primed quantities correspond to the rest frame of the conductor. Let v be the velocity of the conductor, as seen from the magnet frame.

THE ELECTROMAGNETIC SPECTRUM

The electromagnetic (EM) spectrum is the range of all types of EM radiation. Radiation is energy that travels and spreads out as it goes – the visible light that comes from a lamp in your house and the radio waves that come from a radio station are two types of electromagnetic radiation. The other types of EM radiation that make up the electromagnetic spectrum are microwaves, infrared light, ultraviolet light, X-rays and gamma-rays.

You know more about the electromagnetic spectrum than you may think. The image below shows where you might encounter each portion of the EM spectrum in your day-to-day life.

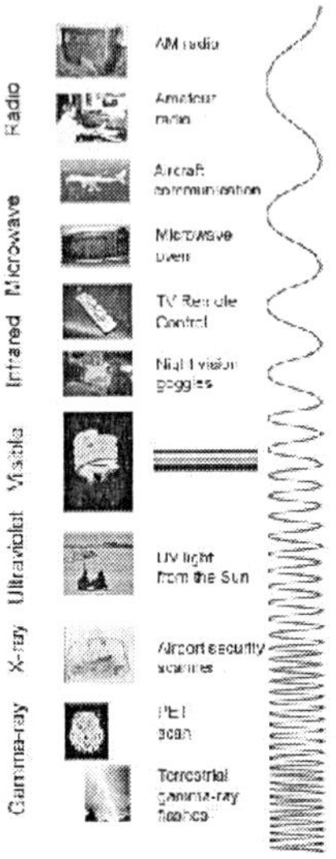

The electromagnetic spectrum from lowest energy/longest wavelength (at the top) to highest energy/shortest wavelength (at the bottom). (Credit: NASA's Imagine the Universe)

Radio: Your radio captures radio waves emitted by radio stations, bringing your favorite tunes. Radio waves are also emitted by stars and gases in space.

Microwave: Microwave radiation will cook your popcorn in just a few minutes, but is also used by astronomers to learn about the structure of nearby galaxies.

Infrared: Night vision goggles pick up the infrared light emitted by our skin and objects with heat. In space, infrared light helps us map the dust between stars.

Visible: Our eyes detect vis ble light. Fireflies, light bulbs, and stars all emit visible light.

Ultraviolet: Ultraviolet radiation is emitted by the Sun and are the reason skin tans and burns. "Hot" objects in space emit UV radiation as well.

X-ray: A dentist uses X-rays to image your teeth, and airport security uses them to see through your bag. Hot gases in the Universe also emit X-rays.

Gamma ray: Doctors use gamma-ray imaging to see inside your body. The biggest gamma-ray generator of all is the Universe.

IS A RADIO WAVE THE SAME AS A GAMMA RAY?

Are radio waves completely different physical objects than gamma-rays? They are produced in different processes and are detected in different ways, but they are not fundamentally different. Radio waves, gamma-rays, visible light, and all the other parts of the electromagnetic spectrum are electromagnetic radiation.

Electromagnetic radiation can be described in terms of a stream of massless particles, called photons, each traveling in a wave-like pattern at the speed of light. Each photon contains a certain amount of energy. The different types of radiation are defined by the the amount of energy found in the photons. Radio waves have photons with low energies, microwave photons have a little more energy than radio waves, infrared photons have still more, then visible, ultraviolet, X-rays, and, the most energetic of all, gamma-rays.

MEASURING ELECTROMAGNETIC RADIATION

Electromagnetic radiation can be expressed in terms of energy, wavelength, or frequency. Frequency is measured in cycles per second, or Hertz. Wavelength is measured in meters. Energy is measured in electron volts. Each of these three quantities for describing EM radiation are related to each other in a precise mathematical way. But why have three ways of describing things, each with a different set of physical units?

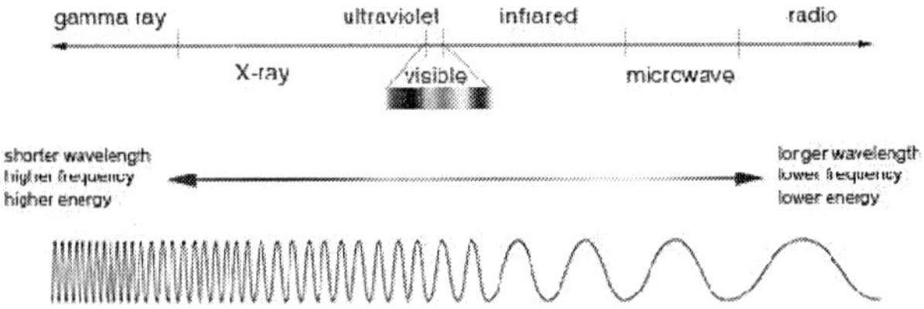

Comparison of wavelength, frequency and energy for the electromagnetic spectrum. (Credit: NASA's Imagine the Universe) The short answer is that scientists don't like to use numbers any bigger or smaller than they have to. It is much easier to say or write "two kilometers" than "two thousand meters." Generally, scientists use whatever units are easiest for the type of EM radiation they work with.

Astronomers who study radio waves tend to use wavelengths or frequencies. Most of the radio part of the EM spectrum falls in the range from about 1 cm to 1 km, which is 30 gigahertz (GHz) to 300 kilohertz (kHz) in frequencies. The radio is a very broad part of the EM spectrum. Infrared and optical astronomers generally use wavelength. Infrared astronomers use microns (millionths of a meter) for wavelengths, so their part of the EM spectrum falls in the range of 1 to 100 microns. Optical astronomers use both angstroms (0.00000001 cm, or 10 cm) and nanometers (0.0000001 cm, or 10 cm). Using nanometers, violet, blue, green, yellow, orange, and red light have wavelengths between 400 and 700 nanometers. (This range is just a tiny part of the entire EM spectrum, so the light our eyes can see is just a little fraction of all the EM radiation around us.) The wavelengths of ultraviolet, X-ray, and gamma-ray regions of the EM spectrum are very small. Instead of using wavelengths, astronomers that study these portions of the EM spectrum usually refer to these photons by their energies, measured in electron volts (eV). Ultraviolet radiation falls in the range from a few electron volts to about 100 eV. X-ray photons have energies in the range 100 eV to 100,000 eV (or 100 keV). Gamma-rays then are all the photons with energies greater than 100 keV. Show me a chart of the wavelength, frequency, and energy regimes of the spectrum

WHY DO WE PUT TELESCOPES IN ORBIT?

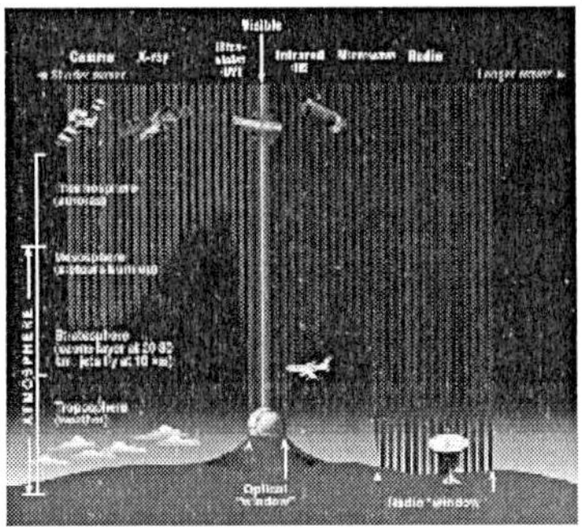

The Earth's atmosphere stops most types of electromagnetic radiation from space from reaching Earth's surface. This illustration shows how far into the atmosphere different parts of the EM spectrum can go before being absorbed. Only portions of radio and visible light reach the surface. (Credit: STScI/JHU/NASA)

Most electromagnetic radiation from space is unable to reach the surface of the Earth. Radio frequencies, visible light and some ultraviolet light makes it to sea level. Astronomers can observe some infrared wavelengths by putting telescopes on mountain tops. Balloon experiments can reach 35 km above the surface and can operate for months. Rocket flights can take instruments all the way above the Earth's atmosphere, but only for a few minutes before they fall back to Earth.

Electromagnetic spectrum, the entire distribution of electromagnetic radiation according to frequency or wavelength. Although all electromagnetic waves travel at the speed of light in a vacuum, they do so at a wide range of frequencies, wavelengths, and photon energies. The electromagnetic spectrum comprises the span of all electromagnetic radiation and consists of many subranges, commonly referred to as portions, such as visible light or ultraviolet radiation. The various portions bear different names based on differences in behaviour in the emission, transmission, and absorption of the corresponding waves and also based on their different practical applications. There are no precise accepted boundaries between any of these contiguous portions, so the ranges tend to overlap.

The entire electromagnetic spectrum, from the lowest to the highest frequency (longest to shortest wavelength), includes all radio waves (e.g., commercial radio and television, microwaves, radar), infrared radiation, visible light, ultraviolet radiation, X-rays, and gamma rays. Nearly all frequencies and wavelengths of electromagnetic radiation can be used for spectroscopy.

DEFINITION OF ELECTROMAGNETIC SPECTRUM

The entire range of wavelengths or frequencies of electromagnetic radiation extending from gamma rays to the longest radio waves and including visible light.

Some Examples of ELECTROMAGNETIC SPECTRUM from the Web

- Such a merger will also give off energy all across the electromagnetic spectrum, from radio waves to gamma rays.
  The Economist, "Gravitational-wave astronomy starts in earnest," 16 Oct. 2017
- Over centuries of telescope construction, astronomers have honed their ability to observe the universe across that electromagnetic spectrum, from radio waves to gamma rays.

Eva Botkin-Kowacki, The Christian Science Monitor, "Nobel Prize for LIGO: How a new era of astronomy will ride on gravitational waves," 3 Oct. 2017

- But the eye is sensitive to only a sliver of the electromagnetic spectrum. Alan Hirshfeld, WSJ, "Catching Waves," 4 Aug. 2017

At 8:32 dave said that waves with higher frequencies contain more energy but microwaves are used to heat food. Heat is a type of energy, so that must mean they contain alot of energy although they have low frequencies. Don't waves with higher frequencies contain more energy?!

Molecules, like water, contain bonds. Bonds are made up of electrons, which are constantly moving. So at temperatures above absolute 0, these bonds are regularly moving, twisting, rotating. Polar molecules, again, like water, contain electric dipoles. This means that the water is partially negatively charged, and the hydrogens are partially positively charged (this occurs because oxygen is more electronegative - a bigger electron hog - than hydrogen).

Microwaves are low energy waves. But when the microwave field passes through, the polar molecules (like water) try to orient themselves in the electric field by moving, twisting, and rotating. As they do, they bump into one another, generating friction, and ultimately, heat.

So the heat energy is not derived directly from the microwaves, but rather from the vibrational movement of polar molecules like water.

Radio waves don't cause any hazards(but they maybe harmful)

microwaves cause Internal heating of the body tissue(which is harmful)

INFRA-RED WAVES CAUSE SKIN BURNS.

The way this is drawn, with B up and E toward us would represent a wave direction going to the left, not to the right. The right hand rule is such that if your fingers are along E and they curl in the direction of B (same as the direction your palm faces) then your thumb sticking out perpendicular points in the direction of the electromagnetic wave, so in your example you'd need to switch the B and E fields drawn to have a wave going to the right.

If x-rays are more energetic than UV rays, then is the reason we are burned by UV rays (sunburn) a function of being exposed to them for longer? Would we receive similar burns from an x-ray if exposed for a similar period of time?

Light: Electromagnetic waves, the electromagnetic spectrum and photons

Properties of electromagnetic radiation and photons.

Google ClassroomFacebookTwitter

Email

INTRODUCTION TO ELECTROMAGNETIC WAVES

Electromagnetic radiation is one of the many ways that energy travels through space. The heat from a burning fire, the light from the sun, the X-rays used by your doctor, as well as the energy used to cook food in a microwave are all forms of electromagnetic radiation. While these forms of energy might seem quite different from one another, they are related in that they all exhibit wavelike properties.

If you've ever gone swimming in the ocean, you are already familiar with waves. Waves are simply disturbances in a particular physical medium or a field, resulting in a vibration or oscillation. The swell of a wave in the ocean, and the subsequent dip that follows, is simply a vibration or oscillation of the water at the ocean's surface. Electromagnetic waves are similar, but they are also distinct in that they actually consist of 222 waves oscillating perpendicular to one another. One of the waves is an oscillating magnetic field; the other is an oscillating electric field.

While it's good to have a basic understanding of what electromagnetic radiation is, most chemists are less interested in the physics behind this type of energy, and are far more interested in how these waves interact with matter. More specifically, chemists study how different forms of electromagnetic radiation interact with atoms and molecules. From these interactions, a chemist can get information about a molecule's structure, as well as the types of chemical bonds it contains. Before we talk about that, however, it's necessary to talk a little bit more about the physical properties of light waves.

Basic properties of waves: Amplitude, wavelength, and frequency

As you might already know, a wave has a trough (lowest point) and a crest (highest point). The vertical distance between the tip of a crest and the wave's central axis is known as its amplitude. This is the property associated with the brightness, or intensity, of the wave. The horizontal distance between two consecutive troughs or crests is known as the wavelength of the wave.

Keep in mind that some waves (including electromagnetic waves) also oscillate in space, and therefore they are oscillating at a given position as time passes. The quantity known as the wave's frequency refers to the number of full wavelengths that pass by a given point in space every second; the SI unit for frequency is Hertz (\text{Hz})(Hz)left parenthesis, H, z, right parenthesis, which is equivalent to "per seconds" \Big((left parenthesiswritten as \dfrac{1}{\text{s}} s 1 start fraction, 1, divided by, s, end fraction or \text{s}^{-1}\Big)s "1)s, start superscript, minus, 1, end superscript, right parenthesis. As you might imagine, wavelength and frequency are inversely proportional: that is, the shorter the wavelength, the higher the frequency, and vice versa. This relationship is given by the following equation:

c=\lambda \nuc=ëí

where \lambdaëlambda (the Greek lambda) is the wavelength (in meters, \text{m}mm) and \nuí (the Greek nu) is the frequency (in Hertz, \text{Hz}HzH, z). Their product is the constant ccc, the speed of light, which is equal to 3.00\times10^8 \text{ m/s}3.00×10 8 m/s3, point, 00, times, 10, start superscript, 8, end superscript, space, m, slash, s. This relationship reflects an important fact: all electromagnetic radiation, regardless of wavelength or frequency, travels at the speed of light.

To illustrate the relationship between frequency and wavelength, let's consider an example.

Example: Calculating the wavelength of a light wave

A particular wave of electromagnetic radiation has a frequency of 1.5\times10^{14}\text{ Hz}1.5×10 14 Hz1, point, 5, times, 10, start superscript, 14, end superscript, space, H, z.

What is the wavelength of this wave?

We can start with our equation that relates frequency, wavelength, and the speed of light.

c=\lambda \nuc=ëí

Next, we rearrange the equation to solve for wavelength.

\lambda=\dfrac{c}{\nu}ë=í c

Lastly, we plug in our given values and solve.

ë=3.00×108ms1.5×10141 s=2.00×10"6 m

Electromagnetic waves can be classified and arranged according to their various wavelengths/frequencies; this classification is known as the electromagnetic spectrum. The following table shows us this spectrum, which consists of all the types of electromagnetic radiation that exist in our universe.

As we can see, the visible spectrum—that is, light that we can see with our eyes—makes up only a small fraction of the different types of radiation that exist. To the right of the visible spectrum, we find the types of energy that are lower in frequency (and thus longer in wavelength) than visible light. These types of energy include infrared (IR) rays (heat waves given off by thermal bodies), microwaves, and radio waves. These types of radiation surround us constantly, and are not harmful, because their frequencies are so low. As we will see in the section, "the photon," lower frequency waves are lower in energy, and thus are not dangerous to our health.

To the left of the visible spectrum, we have ultraviolet (UV) rays, X-rays, and gamma rays. These types of radiation are harmful to living organisms, due to their extremely high frequencies (and thus, high energies). It is for this reason that we wear suntan lotion at the beach (to block the UV rays from the sun) and why an X-ray technician will place a lead shield over us, in order to prevent the X-rays from penetrating anything other than the area of our body being imaged. Gamma rays, being the highest in frequency and energy, are the most damaging. Luckily though, our atmosphere absorbs gamma rays from outer space, thereby protecting us from harm.

Next, we will talk about the relationship between a wave's frequency and its energy.

QUANTIZATION OF ENERGY AND THE DUAL NATURE OF LIGHT

We have already described how light travels through space as a wave. This has been well-known for quite some time; in fact, the Dutch physicist Christiaan Huygens first described the wave nature of light as far back as the late seventeenth century. For about 200200200 years after Huygens, physicists assumed that light waves and matter were quite distinct from one another. According to classical physics, matter was composed of particles that had mass, and whose position in space could be known; light waves, on the other hand, were considered to have zero mass, and their position in space could not be determined. Because they were considered to be in different categories, scientists did not have a good understanding of how light and matter interacted. This all changed in 190019001900, however, when the physicist Max Planck began studying blackbodies – bodies heated until they began to glow.

Planck found that the electromagnetic radiation emitted by blackbodies could not be explained by classical physics, which postulated that matter could absorb or emit any quantity of electromagnetic radiation. Planck observed that matter actually absorbed or emitted energy only in whole-number multiples of the value h\nuh½, where hhh is Planck's constant, 6.626\times10^{-34}\text{ J}\cdot\text{s}6.626×10 "34 JÅ"s6, point, 626, times, 10, start superscript, minus, 34, end superscript, space, J, dot, s, and \nu½ is the frequency of the light absorbed or emitted. This was a shocking discovery, because it challenged the idea that energy was continuous, and could be transferred in any amount. The reality, which Planck discovered, is that energy is not continuous but quantized—meaning that it can only be transferred in individual "packets" (or particles) of the size h\nuh½. Each of these energy packets is known as a quantum (plural: quanta).

While this might sound confusing, we are actually already very familiar with quantized systems. The money we use daily, for example, is quantized. For instance, when you go into a store, you will not see anything on sale for a price of one dollar and two and a half cents (\$1.025)($1.025)left parenthesis, dollar sign, 1, point, 025, right parenthesis. This is because the smallest possible monetary unit is the penny—it is impossible to transfer money in any smaller amount than this. Just as we cannot pay the cashier at the store half of a cent, energy cannot be transferred in anything less than a single quantum. We can think of quanta as being like "pennies" of electromagnetic energy—the smallest possible units by which such energy can be transferred.

Planck's discovery that electromagnetic radiation is quantized forever changed the idea that light behaves purely as a wave. In actuality, light seemed to have both wavelike and particle-like properties.

THE PHOTON

Planck's discoveries paved the way for the discovery of the photon. A photon is the elementary particle, or quantum, of light. As we will soon see, photons can be absorbed or emitted by atoms and molecules. When a photon is absorbed, its energy is transferred to that atom or molecule. Because energy is quantized, the photon's entire energy is transferred (remember that we cannot transfer fractions of quanta, which are the smallest possible individual "energy packets"). The reverse of this process is also true. When an atom or molecule loses energy, it emits a photon that carries an energy exactly equal to the loss in energy of the atom or molecule. This change in energy is directly proportional to the frequency of photon emitted or absorbed. This relationship is given by Planck's famous equation:

E=h\nuE=h½

where EEE is the energy of the photon absorbed or emitted (given in Joules, \text{J}JJ), \nu½is frequency of the photon (given in Hertz, \text{Hz}HzH, z), and hhh is Planck's constant, 6.626\times10^{-34}\text{ J}\cdot\text{s}6.626×10 "34 JÅ"s6, point, 626, times, 10, start superscript, minus, 34, end superscript, space, J, dot, s.

EXAMPLE: CALCULATING THE ENERGY OF A PHOTON

A photon has a frequency of 2.0\times10^{24}\text{ Hz}2.0×10 24 Hz2, point, 0, times, 10, start superscript, 24, end superscript, space, H, z.

What is the energy of this photon?

First, we can apply Planck's equation.

E=h\nuE=h½

Next, we plug in our given value for the frequency, as well as the value for Planck's constant, hhh, and solve.

E=(6.626×10"34 JÅ"s)×(2.0×1024 s"1)=1.3×10"9 J

Concept check: The wavelength of orange light is about 590-635\text{ nm}590"635 nm590, minus, 635, space, n, m, and the wavelength of green light is about 520-560\text{ nm}520"560 nm520, minus, 560, space, n, m. Which color of light is more energetic, orange or green?

(Hint: Keep in mind what you have already learned about the relationship between wavelength and frequency.)

[Show the answer]

CONCLUSION

Electromagnetic radiation can be described by its amplitude (brightness), wavelength, frequency, and period. By the equation E=h\nuE=h?, we have seen how the frequency of a light wave is proportional to its energy. At the beginning of the twentieth century, the discovery that energy is quantized led to the revelation that light is not only a wave, but can also be described as a collection of particles known as photons. Photons carry discrete amounts of energy called quanta. This energy can be transferred to atoms and molecules when photons are absorbed. Atoms and molecules can also lose energy by emitting photons.

So imagine you've got a wave source. This could be a little oscillator that's creating a wave on a string, or a little paddle that goes up and down that creates waves on water, or a speaker that creates sound waves. This could be any wave source whatsoever creates this wave, a nice simple harmonic wave. Now let's say you've got a second wave source. If we take this wave source, the second one, and we put it basically right on top of the first one, we're gonna get wave interference because wave interference happens when two waves overlap. And if we want to know what the total wave's gonna look like we add up the contributions from each wave. So if I put a little backdrop in here and I add the contributions, if the equilibrium point is right here, so that's where the wave would be zero, the total wave can be found by adding up the contributions from each wave. So if we add up the contributions from wave one and wave two wave one here has a value of one unit, wave two has a value of one unit. One unit plus one unit is two units. And then zero units and zero units is still zero. Negative one and negative one is negative two, and you keep doing this and you realize wait, you're just gonna get a really big cosine looking wave. I'm just gonna drop down to here.

We say that these waves are constructively interfering. We call this constructive interference because the two waves combined to construct a wave that was twice as big as the original wave. So when two waves combine and form a wave bigger than they were before, we call it constructive interference. And because these two waves combined perfectly, sometimes you'll hear this as perfectly constructive or totally constructive interference. You could imagine cases where they don't line up exactly correct, but you still might get a bigger wave. In that case, it's still constructive. It might not be totally constructive. So that was constructive interference. And these waves were constructive? Think about it because this wave source two looked exactly like wave source one did, and we just overlapped them and we got double the wave, which is kinda like alright, duh.

That's not that impressive. But check this out. Let's say you had another wave source. A different wave source two. This one is what we call Pi shifted 'cause look at it. Instead of starting at a maximum, this one starts at a minimum compared to what wave source one is at. So it's 1/2 of a cycle ahead of or behind of wave source one. 1/2 of a cycle is Pi because a whole cycle is two Pi. That's why people often call this Pi shifted, or 180 degrees shifted. Either way, it's out of phase from wave source one by 1/2 of a cycle. So what happens if we overlap these two? Now I'm gonna take these two. Let's get rid of that there, let's just overlap these two and see what happens. I'm gonna overlap these two waves. We'll perform the same analysis. I don't even really need the backdrop now because look at. I've got one and negative one. One and negative one, zero.Zero and zero, zero.Negative one and one, zero. Zero and zero, zero and no matter where I'm at, 1/2, a negative 1/2, zero. These two waves are gonna add up to zero. They add up to nothing, so we call this destructive interference because these two waves essentially destroyed each other. This seems crazy. Two waves add up to nothing? How can that be the case? Are there any applications of this? Well yeah. So imagine you're sitting on an airplane and you're listening to the annoying roar of the airplane engine in your ear. It's very loud and it might be annoying. So what do you do? You put on your noise canceling headphones, and what those noise canceling headphones do? They sit on your ear, they listen to the wave coming in. This is what they listen to. This sound wave coming in, and they cancel off that sound by sending in their own sound, but those headphones Pi shift the sound that's going into your ear. So they match that roar of the engine's frequency, but they send in a sound that's Pi shifted so that they cancel and your ear doesn't hear anything. Now it's often now completely silent. They're not perfect, but they work surprisingly well. They're essentially fighting fire with fire. They're fighting sound with more sound, and they rely on this idea of destructive interference. They're not perfectly, totally destructive, but the waves I've drawn here are totally destructive. If they were to perfectly cancel, we'd call that total destructive interference, or perfectly destructive interference. And it happens because this wave we sent in was Pi shifted compared to what the first wave was. So let me show you something interesting if I get rid of all this. Let me clean up this mess. If I've got wave source one, let me get wave source two back. So this was the wave that was identical to wave source one. We overlap 'em, we get constructive interference because the peaks are lining up perfectly with the peaks, and these valleys or troughs are matching up perfectly with the other valleys or troughs. But as I move this wave source too forward, look at what happens. They start getting out of phase.

When they're perfectly lined up we say they're in phase. They're starting to get out of phase, and look at when I move it forward enough what was a constructive situation, becomes destructive. Now all the peaks are lining up with the valleys, they would cancel each other out. And if I move it forward a little more, it lines up perfectly again and you get constructive, move it more I'm gonna get destructive. Keep doing this, I go from constructive to destructive over and over. So in other words, one way to get constructive interference is to take two wave sources that start in phase, and just put them right next to each other. And a way to get destructive is to take two wave sources that are Pi shifted out of phase, and put them right next to each other, and that'll give you destructive 'cause all the peaks match the valleys. But another way to get constructive or destructive is to start with two waves that are in phase, and make sure one wave gets moved forward compared to the other, but how far forward should we move these in order to get constructive and destructive? Well let's just test it out. We start here. When they're right next to each other we get constructive.

If I move this second wave source that was initially in phase all the way to here, I get constructive again. How far did I move it? I moved it this far. The front of that speaker moved this far. So how far was that? Let me get rid of this. That was one wavelength. So look at this picture. From peak to peak is exactly one wavelength. We're assuming these waves have the same wavelength. So notice that essentially what we did, we made it so that the wave from wave source two doesn't have to travel as far to whatever's detecting the sound. Maybe there's an ear here, or some sort of scientific detector detecting the sound. Wave source two is now only traveling this far to get to the detector, whereas wave source one is traveling this far. In other words, we made it so that wave source one has to travel one wavelength further than wave source two does, and that makes it so that they're in phase and you get constructive interference again. But that's not the only option, we can keep moving wave source two forward. We move it all the way to here, we moved it another wavelength forward. We again get constructive interference, and at this point, wave source one is having to make its wave travel two wavelengths further than wave source two does. And you could probably see the pattern. No matter how many wavelengths we move it forward, as long as it's an integer number of wavelengths we again get constructive interference. So something that turns out to be useful is a formula that tells us alright, how much path length difference should there be? So if I'm gonna call this X two, the distance that the wave from wave source two has to travel to get to whatever's detecting that wave. And the distance X one, that wave source one has to travel to get to that detector. So we could write down a formula that relates the difference in path length, I'll call that delta X, which is gonna be the distance that wave one has to travel minus the distance that wave two has to travel. And given what we saw up here, if this path length difference is ever equal to an integer number of wavelengths, so if it was zero that was when they were right next to each other, you got constructive. When this difference is equal to one wavelength, we also got constructive. When it was two wavelengths, we got constructive. It turns out any integer wavelength gives us constructive.

So how would we get destructive interference then? Well let's continue with this wave source that originally started in phase, right? So these two wave sources are starting in phase. How far do I have to move it to get destructive? Well let's just see. I have to move it 'til it's right about here. So how far did the front of that speaker move? It moved about this far, which if I get rid of that speaker you could see is about 1/2 of a wavelength. From peak to valley, is 1/2 of a wavelength, but that's not the only option. I can keep moving it forward. Let's just see, that's constructive. My next destructive happens here which was an extra this far. How far was that? Let's just see. That's one wavelength, so notice at this point, wave source one is having to go one and 1/2 wavelengths further than wave source two does. So let's just keep going.

Move wave source two, that's constructive. We get another destructive here which is an extra this far forward, and that's equal to one more wavelength. So if we get rid of this you could see valley to valley is a whole nother wavelength. So in this case, wave source two has to travel two and 1/2 wavelengths farther than wave source two. Any time wave source one has to travel 1/2 integer more wavelengths than wave source two, you get destructive interference. In other words, if this path length difference here is equal to lambda over two, three lambda over two, which is one and 1/2 wavelengths. Five lambda over two, which is two and 1/2 wavelengths, and so on, that leads to destructive interference. So this is how the path length differences between two wave sources can determine whether you're gonna get constructive or destructive interference. But notice we started with two wave sources that were in phase. These started in phase. So this whole analysis down here assumes that the two sources started in phase with each other, i.e. neither of them are Pi shifted. What would this analysis give you if we started with one that was Pi shifted? So let's get rid of this wave two. Let's put this wave two back in here. Remember this one? This one was Pi shifted relative to relative to wave source one. So if we put this one in here, and we'll get rid of this, now when these two wave sources are right next to each other you're getting destructive interference. So this time for a path length difference of zero, right? These are both traveling the same distance to get to the detector. So X one and X two are gonna be equal. You subtract them, you'd get zero. This time the zero's giving us destructive instead of constructive. So let's see what happens if we move this forward, let's see how far we've gotta move this forward to again get destructive.

We'd have to move it over to here. How far did we move it? Let's just check. We moved the front of this speaker that far, which is one whole wavelength. So if we get rid of this, we had to move the front of the speaker one whole wavelength, and look at again it's destructive. So again, zero gave us destructive this time, and the lambda's giving us destructive, and you realize oh wait, all of these integer wavelengths. If I move it another integer wavelength forward, I'm again gonna get destructive interference because all these peaks are lining up with valleys. So interestingly, if two sourcese started Pi out of phase, so I'm gonna change this. Started Pi out of phase, then path length differences of zero, lambda, and two lambda aren't gonna give us constructive, they're gonna give us destructive. And so you could probably guess now, what are these path length differences of 1/2 integer wavelengths gonna give us? Well let's just find out. Let's start here, and we'll get rid of these. Let's just check. We'll move this forward 1/2 of a wavelength and what do I get? Yup, I get constructive.

So if I move this Pi shifted source 1/2 a wavelength forward instead of giving me destructive, it's giving me constructive now. And if I move it so it goes another wavelength forward over to here, notice this time wave source one has to move one and 1/2 wavelengths further than wave source two. That's 3/2 wavelengths. But instead of giving us destructive, look. These are lining up perfectly. It's giving us constructive, and you realize oh, all these 1/2 integer wavelength path length differences, instead of giving me destructive are giving me constructive now because one of these wave sources was Pi shifted compared to the other. So I can take this here, and I could say that when the two sources start Pi out of phase, instead of leading to destructive this is gonna lead to constructive interference. And these two ideas are the foundation of almost all interference patterns you find in the universe, which is kind of cool. If there's an interference pattern you see out there, it's probably due to this. And if there's an equation you end up using, it's probably fundamentally based on this idea if it's got wave interference in it. So I should say one more thing, that sources don't actually have to start out of phase. Sometimes they travel around. Things happen, it's a crazy universe. Maybe one of the waves get shifted along its travel. Regardless, if any of them get a Pi shift either at the beginning or later on, you would use this second condition over here to figure out whether you get constructive or destructive. If neither of them get a phase shift, or interestingly, if both of them get a phase shift, you could use this one 'cause you could imagine flipping both of them over, and it's the same as not flipping any of them over. So recapping, constructive interference happens when two waves are lined up perfectly. Destructive interference happens when the peaks match the valleys and they cancel perfectly. And you could use the path length difference for two wave sources to determine whether those waves are gonna interfere constructively or destructively. The path length difference is the difference between how far one wave has to travel to get to a detector compared to how far another wave has to travel to get to that same detector, assuming those two sources started in phase and neither of them got a Pi shift along their travels. Path length differences of integer wavelengths are gonna give you constructive interference, and path length differences of 1/2 integer wavelengths are gonna give you destructive interference. Whereas if the two sources started Pi out of phase, or one of the got a Pi phase shift along its travel, integer wavelengths for the path length difference are gonna give you destructive interference. And 1/2 integer wavelengths for the path length difference are gonna give you constructive interference.

6

EMF for non-thin-wire Circuits

Suppose we have a single open circuit wire, moving with speed vv inside an infinite and uniform magnetic field BB. Wire has a length of LL and it moves perpendicular to magnetic field. In this case Lorentz force is used to calculate the EMF created in wire and it is:

EMF=vBLEMF=vBL

We know that this formula is derived using Lorentz force and also Lorentz force can be deduced from Maxwell-Faraday equation (explained here), so can we obtain the same above formula using Maxwell-Faraday equation?

cE.dl="ddt,"sB.ds."cE.dl="ddt,"sB.ds

If yes, how ? Also here says that Faraday's law of induction can not be applied to a single wire:

The induced electromotive force in any closed circuit is equal to the negative of the time rate of change of the magnetic flux enclosed by the circuit.

This version of Faraday's law strictly holds only when the closed circuit is a loop of infinitely thin wire, and is invalid in other circumstances.

Electric Field of Single-Battery Circuit with Thin/Thick Wires Mar 24, 2010 #1 DeadFishFactory 1. The problem statement, all variables and given/ known data The circuit shown above consists of a single battery, whose emf is 1.4 V, and three wires made of the same material, but having different cross-sectional areas. Each thick wire has cross-sectional area 1.4e-6 m2, and is 21 cm long. The thin wire has cross-sectional area 6.4e-8 m2, and is 7.5 cm long. In this metal, the electron mobility is 6e-4 (m/s)/(V/m), and there are 8e+28 mobile electrons/m3. Use the appropriate equation(s), plus the equation relating electron current to electric field, to solve for the factor that goes in the blank below: EF = ? * ED Use the appropriate equation(s) to calculate the magnitude of ED ED = ? Use the appropriate equation(s) to calculate the electron current at location D in the steady state 2. Relevant equations ??? 3. The attempt at a solution Already tried it, and I can't do it. E = emf/L doesn't work here. I've looked everywhere to try and find equations that can help me solve it, or at least relate the thin and thick wire. My physics book seems to skip this whole section (I'm using a different physics book than my class, which kind of sux for me, but I didn't want to buy another one).

I solved it all on my own. I'll post the solution here so people will know how to do it if they come across it. i = naME i = electron current n = electron density a = area M = electron mobility E = electric field For the thin and thick wire, the electron current for both of them is constant. So you can put the two together. nA(thick)ME(thick) = nA(thin)ME(thin) A(thick)E(thick) = A(thin)E(thin), n and M are constants, too. E(thick) = [A(thin)/A(thick)]E(thin) A(thin) = 6.4E-8 m^2 A(thick) = 1.4E-6 m^2 E(thick) = 0.046E(thin) You have the equation: 0 = 1.4 - EF*0.21 - ED*0.075 - EF*0.21 [Found by applying the Loop Rule] E(thick) = 0.046E(thin). So, EF = E(thick) and ED = E(thin) from the diagram. Plug it in. 0 = 1.4 - 2(0.21)(0.046)E(thin) E(thin) = 14.8 V/M You have i=naME to find i. So plug that in. M = 6E-4 n = 8E28 A(thin) = 6.4E-8 E = 14.8 i = (6E-4)(8E28)(6.4E-8)(14.8) = 4.5E19 electrons / second.

EMF FOR NON-THIN-WIRE CIRCUITS

It is tempting to generalize Faraday's law to state that If "Ó is any arbitrary closed loop in space whatsoever, then the total time derivative of magnetic flux through Ó equals the EMF around "Ó. This statement, however, is not always true—and not just for the obvious reason that EMF is undefined in empty space when no conductor is present. As noted in the previous section, Faraday's law is not guaranteed to work unless the velocity of the abstract curve "Ó matches the actual velocity of the material conducting the electricity. The two examples illustrated below show that one often obtains incorrect results when the motion of "Ó is divorced from the motion of the material.

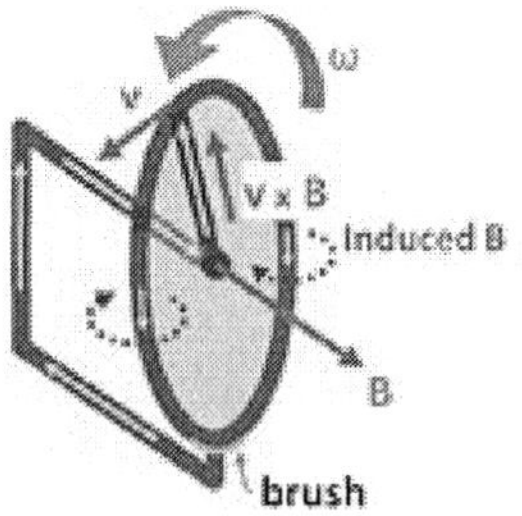

Faraday's homopolar generator. The disc rotates with angular rate ù, sweeping the conducting radius circularly in the static magnetic field B. The magnetic Lorentz force v × B drives the current along the conducting radius to the conducting rim, and from there the circuit completes through the lower brush and the axle supporting the disc. This device generates an EMF and a current, although the shape of the "circuit" is constant and thus the flux through the circuit does not change with time.

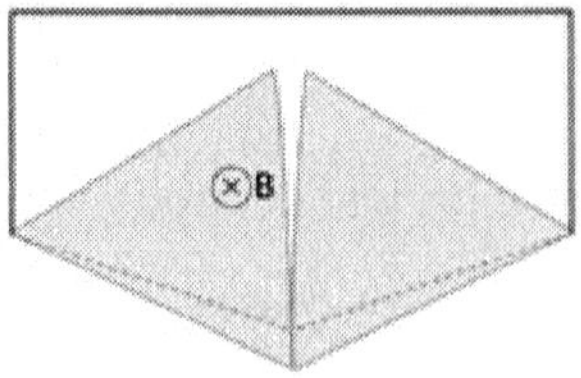

A wire (solid red lines) connects to two touching metal plates (silver) to form a circuit. The whole system sits in a uniform magnetic field, normal to the page. If the abstract path "Ó follows the primary path of current flow (marked in red), then the magnetic flux through this path changes dramatically as the plates are rotated, yet the EMF is almost zero. After Feynman Lectures on Physics Vol. II page 17-3.

MAGNETOSTATICS

Magnetostatics is the study of magnetic fields in systems where the currents are steady (not changing with time). It is the magnetic analogue of electrostatics, where the charges are stationary. The magnetization need not be static; the equations of magnetostatics can be used to predict fast magnetic switching events that occur on time scales of nanoseconds or less. Magnetostatics is even a good approximation when the currents are not static — as long as the currents do not alternate rapidly. Magnetostatics is widely used in applications of micromagnetics such as models of magnetic recording devices. Magnetostatic focussing can be achieved either by a permanent magnet or by passing current through a coil of wire whose axis coincides with the beam axis.

MAGNETOSTATICS AS A SPECIAL CASE OF MAXWELL'S EQUATIONS

Starting from Maxwell's equations and assuming that charges are either fixed or move as a steady current {\displaystyle \scriptstyle \mathbf {J} }, the equations separate into two equations for the electric field (see electrostatics) and two for the magnetic field. The fields are independent of time and each other. The magnetostatic equations, in both differential and integral forms:

Where " denotes divergence, and B is the magnetic flux density, the first integral is over a surface {\displaystyle \scriptstyle S} with oriented surface element {\displaystyle \scriptstyle d\mathbf {S} }. Where J is the current density and H is the magnetic field intensity, the second integral is a line integral around a closed loop {\displaystyle \scriptstyle C} with line element {\displaystyle \scriptstyle \mathbf {l} }. The current going through the loop is {\displaystyle \scriptstyle I_{\text{enc}}}.

The quality of this approximation may be guessed by comparing the above equations with the full version of Maxwell's equations and considering the importance of the terms that have been removed. Of particular significance is the comparison of the {\displaystyle \scriptstyle \mathbf {J} } term against the {\displaystyle \scriptstyle \partial \mathbf {D} /\partial t} term. If the {\displaystyle \scriptstyle \mathbf {J} } term is substantially larger, then the smaller term may be ignored without significant loss of accuracy.

RE-INTRODUCING FARADAY'S LAW

A common technique is to solve a series of magnetostatic problems at incremental time steps and then use these solutions to approximate the term {\displaystyle \scriptstyle \partial \mathbf {B} /\partial t}. Plugging this result into Faraday's Law finds a value for {\displaystyle \scriptstyle \mathbf {E} } (which had previously been ignored). This method is not a true solution of Maxwell's equations but can provide a good approximation for slowly changing fields.

SOLVING FOR THE MAGNETIC FIELD

CURRENT SOURCES

This technique works well for problems where the medium is a vacuum or air or some similar material with a relative permeability of 1. This includes air-core inductors and air-core transformers. One advantage of this technique is that, if a coil has a complex geometry, it can be divided into sections and the integral evaluated for each section. Since this equation is primarily used to solve linear problems, the contributions can be added. For a very difficult geometry, numerical integration may be used.

For problems where the dominant magnetic material is a highly permeable magnetic core with relatively small air gaps, a magnetic circuit approach is useful. When the air gaps are large in comparison to the magnetic circuit length, fringing becomes significant and usually requires a finite element calculation. The finite element calculation uses a modified form of the magnetostatic equations above in order to calculate magnetic potential. The value of {\displaystyle \scriptstyle \mathbf {B} } can be found from the magnetic potential.

The magnetic field can be derived from the vector potential. Since the divergence of the magnetic flux density is always zero, {\displaystyle \mathbf {B} =\nabla \times \mathbf {A}, }and the relation of the vector potential to current.

Strongly magnetic materials (i.e., ferromagnetic, ferrimagnetic or paramagnetic) have a magnetization that is primarily due to electron spin. In such materials the magnetization must be explicitly included using the relation{\displaystyle \mathbf {B} =\mu _{0}(\mathbf {M} +\mathbf {H}).}

Except in metals, electric currents can be ignored. Then Ampère's law is simply The vector potential method can also be employed with an effective current density. Since an E.M.F is induced in a rotating copper disc placed in a constant magnetic field, can Faraday's law be applied here?

If you're talking about the situation I suspect, a Homopolar generator, where the rotation axis is parallel to the magnetic field, then indeed Faraday's law doesn't strictly apply, at least not usefully, because the magnetic flux is constant over any area of the disk, so the time rate of change of flux and thus the EMF around the edge of the area is zero. You should really be applying the Lorentz force law, in conjunction with the fact that rotating the disk tends to drag the electrons in it tangentially, so that the magnetic field deflects them radially.

However, it turns out that a naive application of Faraday's law gives the same answer because of a combination of two things: (i) the relativity principle, which means that there's a precise correspondence between Faraday's law in one frame and the Lorentz force in another, and (ii) the symmetry of the situation, which means you get away with applying the relativity principle even though it doesn't strictly apply to rotating frames.{\displaystyle \mathbf {J_{M}} =\nabla \times \mathbf {M}. }

This depends on how the disc rotates. I'll assume for simplicity that the disk rotates such that the area seen by the magnetic field changes. Otherwise, the induced EMF is zero magnetic field doesn't change, area doesn't change. Let me just present you with a simple example (common) that you see with Faraday's law so you understand about this change in area.

For your particular question, let's approach it formally. What is the total area through with the magnetic field passes through at a given time?

A(t)=Acos(ùt+ä)A(t)=Acosa (ùt+ä)

Depending on how we initialize the system A(0)=?A(0)=?, that gives us our value for ää. Given the magnetic field BB - it appears we have everything we need to know for this question:

As the loop rotates around, the total area seen by the magnetic field changes, which causes e.m.f. as there is a change in flux (less field lines poke through the surface) and Nature Abhors A Change In Flux kicks in.

Faraday actually discovered this phenomenon in his original experiments on electricity and magnetism. It has come to be called a "homopolar generator." It was also called "Faraday's Electromagnetic Paradox" because it violates what we call Faraday's Law, and Faraday himself could not explain it. It is correctly explained by the Lorenz Force.

Due to rotation e-s experience a magnetic force and different polarity occurs in different sides. And a reverse electric fields appears. Now electrostatic force.isequal to magnetic force(lorrentz). $-qE= -q(v\times B)$

You can view the induced emf as motional emf. It is similar to the case when a conducting rod is moved through a uniform magnetic field perpendicular to it. This induces an emf, $E=Blv$ in the rod.

Actually, during the motion of the rod, the electrons experience a magnetic force and tend to accumulate on the ends of the rod. This causes an emf to be induced in the opposite direction to the magnetic force.

Finally, in steady state, the two forces become equal. Let E be the final emf, Electric field = E/l

In steady state, Fm = Fe or, evB = eE/l=> E = Bvl

The case of the disc can be considered as many radial conducting rods arranged together,each having the same emf. but this time the velocity of each point of the rod is different. (use v= wr) and integrate to get E=1/2 Bwr^2

Whenever there is a change in magnetic feild,induced emf will be produced. we can change magnetic feild in two types.

1. By supplying alternating current as flux varies with current there is a change in flux.i.e. transformer principle (faradays day)
2, `we can cut constant magnetic feild by a conductor which makes flux to change.

Induced emf =Bvl..for conductor moving with velocity v.i.e generator principle by changing the effective area of conductor loop emf=BAw.

Thus, Yes off coarse because for induction of an emf In a coil either of two must be there 1.alternating emf or2.motion of object placed in static magnetic field....

One piece Faraday generator: A paradoxical experiment from 1851

ABSTRACT

In the conventional Faraday generator a conducting disk rotates in an axial magnetic field. If the disk is replaced by a cylindrical permanent magnet that supplies its own magnetic field, the effect is identical. It follows that any moving magnet generates an induced electromotive force due to the presence of its own field: this generalization leads to an apparent paradox in the case of translational motion for it implies the possibility that an observer in an inertial frame could measure his absolute velocity.

Teaching Faraday's law of electromagnetic induction in introductory physics courses is challenging. We discuss some inaccuracies in describing a moving conductor in the context of electromagnetic induction. Among them is the use of the ambiguous term "area change" and the unclear relation between Faraday's law and Maxwell's equation for the electric field circulation. We advocate the use of an expression for Faraday's law that shows explicitly the contribution of the time variation of the magnetic field and the action of the Lorentz force, which are usually taught separately. This expression may help students' understanding of Faraday's law and lead to improved problem solving skills.

Any change in the magnetic environment of a coil of wire will cause a voltage (emf) to be "induced" in the coil. No matter how the change is produced, the voltage will be generated. The change could be produced by changing the magnetic field strength, moving a magnet toward or away from the coil, moving the coil into or out of the magnetic field, rotating the coil relative to the magnet, etc.

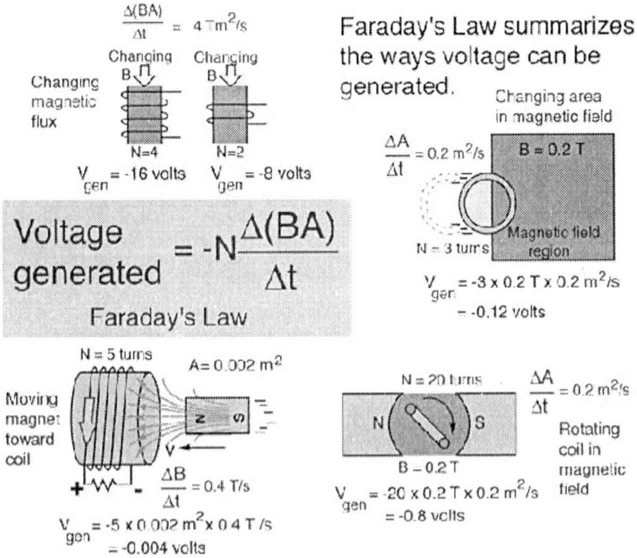

Further comments on these examples

Faraday's law is a fundamental relationship which comes from Maxwell's equations. It serves as a succinct summary of the ways a voltage (or emf) may be generated by a changing magnetic environment. The induced emf in a coil is equal to the negative of the rate of change of magnetic flux times the number of turns in the coil. It involves the interaction of charge with magnetic field.

LENZ'S LAW

When an emf is generated by a change in magnetic flux according to Faraday's Law, the polarity of the induced emf is such that it produces a current whose magnetic field opposes the change which produces it. The induced magnetic field inside any loop of wire always acts to keep the magnetic flux in the loop constant. In the examples below, if the B field is increasing, the induced field acts in opposition to it. If it is decreasing, the induced field acts in the direction of the applied field to try to keep it constant.

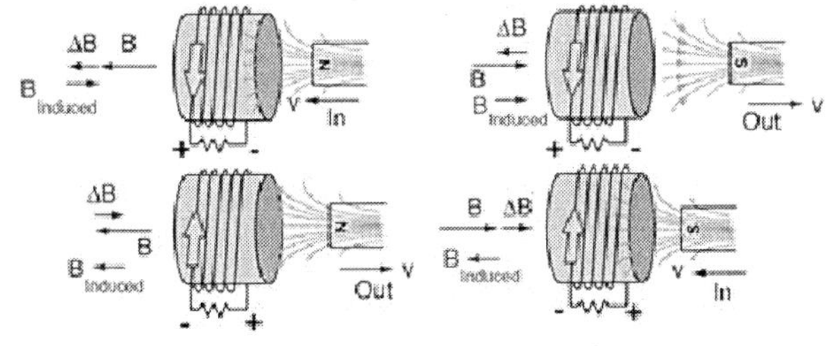

MAGNET AND COIL

When a magnet is moved into a coil of wire, changing the magnetic field and magnetic fluxthrough the coil, a voltage will be generated in the coil according to Faraday's Law. In the example shown below, when the magnet is moved into the coil the galvanometer deflects to the left in response to the increasing field. When the magnet is pulled back out, the galvanometer deflects to the right in response to the decreasing field. The polarity of the induced emf is such that it produces a current whose magnetic field opposes the change that produces it. The induced magnetic field inside any loop of wire always acts to keep the magnetic flux in the loop constant. This inherent behavior of generated magnetic fields is summarized in Lenz's Law.

Failling to understand why a rotating disk inside an uniform magnetic field generates a potential difference

Suppose we have a conducting disk of radius RR, rotating about its axis, with rotational velocity ÉÉ. If surrounding the disk is a constant magnetic field parallel to its normal vector, a non-zero voltage appears between the center and the border of the disk.

I believe the answer should be V=%"B%" A fV= B A f, with f=É2Àf=É2À, but I have no idea where this comes from, or why what I did yields an incorrect answer.

Also, could anyone illustrate the differences between EE, the induced EMF, and VV here? They always seem to be the same thing...

What bjorne said is right but incomplete. The effect is not due to the flux.

Due to Lorentz force electrons which are moving at a distance rr from the center with tangential velocity v=rÉv=rÉ (due to rotation of the disk) experience a radial force (do the vector product to see this)

FL=erÉBFL=erÉB

WHERE B IS THE FIELD AND EE THE ELECTRONS CHARGE.

This leads to electrons moving towards the periphery and creates a charge discontinuity (also because they leave their nuceli behind) which generates a radial electric field (it is radial because it is due to a radial gradient of charge) E(r)=dVdrE(r)=dVdr and thus a voltage V(r)V(r). The fact that E=dVdrE=dVdr is due to the symmetry of the problem: all other components of the gradient vanish as the field only variates in the radial component.

In the steady state the electric (FE=eE=edVdrFE=eE=edVdr) and Lorentz forces cancel out as nothing should move at equilibrium.

So

erÉB=edVdrerÉB=edVdr

and this gives us an equation for the voltage which when integrated leads to:

V(r)=BÉr22V(r)=BÉr22

with respect to the center of the disk where V=0V=0.

Between the edge of the disk the voltage is

V(R)=BÉR22V(R)=BÉR22

WHICH IS THE SAME FORMULA YOU PROPOSE.

EDIT: Notice that in this answer I have been neglecting centrifugal force FC=meÉ2rFC=meÉ2r acting on the electrons. We can add it to show that the effect, the fact that a voltage arises, is not a purely magnetical effect.

Once again the trick is saying that in the steady state nothing moves, so the total force must be zero. That means:

edVdr=erÉB+meÉ2redVdr=erÉB+meÉ2r

or rather

dVdr=(B+meeÉ)ÉrdVdr=(B+meeÉ)Ér

which when integrated leads to

V(r)=(B+meeÉ)Ér22V(r)=(B+meeÉ)Ér22

Thus even if B=0B=0 we still get a voltage V(R)=meeÉ2R22V (R)=mee É2R22 across the disk.

Yet notice that the effect is small as for an electron mee<" 10"11mee <"10"11 (which by the way is in principle measurable from this device).

- Hmm, I'm beginning to understand this, I think. I have a few doubts though: The force an electron in the disk feels should be F=e (E+v×B)F=e (E+v×B), right? It's easy to see, as you said, the FmFm, the magnetic force, is radial and of magnitud %"Fm%"=eÉrB Fm =eÉrB. Now, I don't really see why the electric field generated by the displacement of the electrons by the previous force is radial (and thus, why the electric force is radial). – YoTengoUnLCD Nov 6 '16 at 22:21
- Given that, it's clear that in the steady state we should have %"E%"=%"v×B%"=ÉrB E = v×B =ÉrB. However, this leads to %""V%"=ÉrB "V =ÉrB (as ""V=E""V=E). How does this lead

to V=Ér22BV=Ér22B (It's a bit unclear why you can write %"E%"=dVdr E =dVdr)? – YoTengoUnLCD Nov 6 '16 at 22:21

- @YoTengoUnLCD, your set up leads to only radial fields/forces/voltages since there is symmetry around the axis. There's no reason there should be force is anything other than the radial direction. – bjorne Nov 6 '16 at 22:23
- Symmetry. The displacement of the electron is radial and the electic field is born to contrast this displacement so it is radial too. Thus dV/dr is the only component of the gradient left, he others being constant. By integration you get the final answer. – JalfredP Nov 7 '16 at 8:42
- I edited to clarify and added a sub-part about the effect of centrifugal force.

I think the formula you're using is E="d¦/dtE="d¦/dt where ¦¦ is the magnetic flux through your conducting loop, or in your case, conducting material. Without the algebra you can tell that the disk is always in the same magnetic field, so the flux through it is unchanging. The motional emf is then indeed zero. The force must be coming from somewhere else - not Faraday's law of induction.The part of electromagnetism that explains this phenomenon is the Lorentz force. A charge moving with velocity vv in a magnetic field BB is acted upon by a force per unit charge of v×Bv×B. From the set up, vv is always tangential to the disk and of magnitude ÉrÉr, while BB is always vertical, so the force will always be radial. The conducting disk has positively charged nucleii that are stuck in the lattice, but the electrons subject to the force will move and induce a current (like in a homopolar dynamo), or "pile up" with a voltage between different radii of the disk.A steady state will quickly be reached where the Lorentz force is balanced by the electric field, which allows us to calculate the voltage between the centre and a point of radius RR (including the edge),V(R)="+" R0EÅ"dl=+" R0BÉr dr=BÉR2/2

I think that in the original question there is a confusion between e.m.f. EE and (electrostatic) Voltage VV. As others have pointed out, if enough time passes a static radial electric field E× E'! will appear due to Lorentz forces acting on the mobile conducting electrons, caused by a charge separation in the disk. In this limit we have an electrostatic situation, the field E× E'! can be considered the gradient of a potencial and the potencial difference is as shown above.However, if a circuit is closed by connecting a point PP in the perifery of the disk to its conducting axis via a wire with resistance RR, then the radius OPOP can be seen as a moving segment of the closed circuit, and as it moves it changes the circuit area at a rate dAdt= 12ÉR2dAdt= 12ÉR2 (assuming everything else remains in place) and thus the flux rate is indeedE="BdAdt="12ÉR2B.E="BdAdt="12ÉR2B.This e.m.f will induce a current i=E/Ri=E/R in the circuit (assume the disk is a perfect conductor) but we cannot associate an electrostatic field E× E'! to each point in the disk (there is no charge separation now), although a tension can de defined between PP and OO as VPO="EVPO="E by integration of the stationary electric field in the wire.

Induced EMF Almost 200 years ago, Faraday looked for evidence that a magnetic field would induce an electric current with t Induced EMF

He found no evidence of currents in Y when the current in X was steady, but did see a current induced when the switch was turned on or off. He also observed a current when a magnet was moved into or out of a loop, but not if it was kept stationary.

Therefore, the appropriate conclusion is that a changing magnetic field induces an emf. Faraday's experiment used a magnetic field that was changing because the current producing it was changing; the previous graphic shows a magnetic field that is changing because the magnet is moving. Either type of change produces a current.

More precise statement of experimental results: The induced emf in a wire loop is proportional to the rate of change of magnetic flux through the loop. Magnetic flux: Unit of magnetic flux: weber, Wb. 1 Wb = 1 T?m2

This drawing shows the variables in the defining equation for magnetic flux:

The magnetic flux is analogous to the electric flux - it is proportional to the total number of field lines passing through the loop.

The minus sign in Faraday's Law gives the direction of the induced emf: Lenz's Law gives a different statement related to the direction of an induced current: A current produced by an induced emf moves in a direction so that the magnetic field it produces ten

to restore the changed magnetic flux. Often stated slightly differently, with same meaning.

MAGNETIC FLUX WILL CHANGE IF THE AREA OF THE LOOP CHANGES:

Magnetic flux will change if the angle between the loop and the field changes:

- Determine whether the magnetic flux is increasing, decreasing, or unchanged.
- The magnetic field due to the induced current points in the opposite direction to the original field if the flux is increasing; in the same direction if it is decreasing; and is zero if the flux is not changing.
- Use the righthand rule to determine the direction of the current.
- Remember that the external field and the field due to the induced current are describing different contributions.

EMF Induced in a Moving Conductor The induced current is in a direction that tends to slow the moving bar - it will take an external force to keep it moving. Why ? Think magnetic force on charge due to the induced motion of the charges.

EMF INDUCED IN A MOVING CONDUCTOR THE INDUCED EMF HAS MAGNITUDE

Changing Magnetic Flux Produces an Electric Field A changing magnetic flux induces an electric field; this is a generalization of Faraday's law.

The electric field will exist regardless of whether or not there are any conductors around to carry an induced current.This is a different way of looking at induced emf- it creates an electric field recall the relationship between E and ?V

he relationship between E and ?V Electric Generators A generator is the opposite of a motor - it transforms mechanical energy into electricalenergy. This is an ac generator: The axle is rotated by an external force such as falling water or steam. The brushes are in constant electrical contact with the slip rings.

Electric Generators A sinusoidal emf is induced in the rotating loop (N is the number of turns, A is area of the loop, and ? is 2? f).

Transformers This is a stepup transformer - the emf in the secondary coil is larger than the emf in the primary. AC current in primary coil creates an AC Bfield in the soft iron core which then induces an emf (VS) in the secondary coil. ?? is the same in one turn of each winding

Transformers A transformer consists of two coils, either interwoven or linked by an iron core. A changing emf in one induces an emf in the other. The ratio of the emfs is equal to the ratio of the number of turns in each coil when wound on a common core (same ?B in single loop of each)

Mutual inductance: a changing current in one coil will induce a current in a second coil.

And vice versa; note that the constant M, known as the mutual inductance.

Energy Stored in a Magnetic Field * Previously, we saw that energy can be stored in an electric field, energy can be stored in a magnetic field as well, in an inductor.

MAXWELL'S EQUATIONS (INTEGRAL FORM)

It is sometimes easier to understand Maxwell's equations in their integral form; the version we outlined last time is the differential form.

For Gauss' law and Gauss' law for magnetism, we've actually already done this. First, we write them in differential form:

$$\nabla \cdot E = \frac{1}{\epsilon_0}\rho$$
$$\nabla \cdot B = 0$$

We pick any region V we want and integrate both sides of each equation over that region:

On the left-hand sides we can use the divergence theorem, while the right sides can simply be evaluated:

where $Q(V)$ is the total charge contained within the region . Gauss' law tells us that the flux of the electric field out through a closed surface is (basically) equal to the charge contained inside the surface, while Gauss' law for magnetism tells us that there is no such thing as a magnetic charge. Faraday's law was basically given to us in integral form, but we can get it back from the differential form:

$$\nabla \times E = -\frac{\partial B}{\partial t}$$

We pick any surface and integrate the flux of both sides through it:

$$\int_S \nabla \times E \cdot dS = \int_S -\frac{\partial B}{\partial t} \cdot dS$$

On the left we can use Stokes' theorem, while on the right we can pull the derivative outside the integral:

$$\int_{\partial S} E \cdot dr = -\frac{\partial}{\partial t}\Phi_S(B)$$

where $\Phi_S(B)$ is the flux of the magnetic field B through the surface S. Faraday's law tells us that a changing magnetic field induces a current around a circuit.

A SIMILAR ANALYSIS HELPS WITH AMPÈRE'S LAW:

$$\nabla \times B = \mu_0 J + \epsilon_0\mu_0\frac{\partial E}{\partial t}$$

WE PICK A SURFACE AND INTEGRATE:

$$\int_S \nabla \times B \cdot dS = \int_S \mu_0 J \cdot dS + \int_S \epsilon_0\mu_0\frac{\partial E}{\partial t} \cdot dS$$

THEN WE SIMPLIFY EACH SIDE

$$\int_{\partial S} B \cdot dr = \mu_0 I_S + \epsilon_0\mu_0\frac{\partial}{\partial t}\Phi_S(E)$$

where $\Phi_S(E)$ is the flux of the electric field E through the surface S, and I_S is the total current flowing through the surface . Ampère's law tells us that a flowing current induces a magnetic field around the current, and Maxwell's correction tells us that a changing electric field behaves just like a current made of moving charges.

- We collect these together into the integral form of Maxwell's equations:
- Physics Project Report on Electromagnetic Induction
- Physics Project Report on Electromagnetic Induction :
- Electromagnetic Induction (EMI) : In 1831 Michael Faraday discovered the effect called "Electromagnetic Induction" just converse to the magnetic effect of electric current.

When a coil made of copper wire is placed inside a magnetic field, magnetic flux is linked with the coil. Faraday found that when the magnetic flux linked with the coil is changed, an electric current starts flowing in the coil, provided the coil is closed one. The current and e.m.f. so produced are called induced current and induced e.m.f. The induced current and the e.m.f. in the coil last only so long as the magnetic flux linked with the coil keeps on changing.

Thus electromagnetic induction is the phenomenon of production of electric current (or e.m.f.) in a coil when the magnetic flux linked with the coil is changed.

FARADAY'S EXPERIMENT

The following experiment performed by Faraday led to the discovery of the electromagnetic induction.

WHEN THE STRENGTH OF MAGNETIC FIELD IS VARIED :

Consider two coil P and S wound on an iron rod. Iron rod is connected with galvanometer, battery and tapping key. When tapping key is pressed and when it is released galvanometer shows deflection showing the presence of induced current.

Explanation : When the tapping key is pressed then magnetic flux linked with the coil S changed because of increase in magnetic field of coil P and induced current is produced and when it is released magnetic flux is again changed and induced current is produced. But when the tapping key is kept pressed then the magnetic flux linked with coil do not changed and induced current do not produce so galvanometer shows no deflection.

Faraday's Laws of Electromagnetic Induction :

The results of Faraday's experiment on electromagnetic induction are known as "Faraday's Law of Electromagnetic Induction". These are stated as below :

1. Whenever magnetic flux linked with a circuit (a loop of wire or a coil or an electric circuit in general) changes, induced e.m.f. is produced.
2. The induced e.m.f. lasts as long as the change in magnetic flux continuous.
3. The magnitude of induced e.m.f. is directly proportional to the rate of change of magnetic flux linked with the circuit.

LENZ'S RULE :

Lenz's rule is a convenient method to determine the direction of induced current produced in the circuit.

Lenz's law states that the induced current produced in a circuit always flows in such a direction that it opposes the change or cause that produce it.

Let us now apply Lenz's law to find the direction of flow of induced current in the circuit.

On pressing the key the current in the coil P flows in clockwise direction and magnetic lines of force are directed from left to right. Then magnetic flux linked with the coil S changed. The direction of induced current should be such that it should oppose the direction of flow of magnetic field lines. So induced current in the coil S is in the direction opposite to the magnetic field in P. Hence, direction of induced current in coil S is from right to left. So induced current in coil S should flow in anticlockwise direction.

EXPRESSION FOR MOTIONAL E.M.F. :

Consider that a uniform magnetic field B confined to the region PQRS and a coil ABCD is placed inside the magnetic field. The direction of magnetic field is perpendicular to the plane of the coil and in inward direction.

Consider that at any time t, the part BA′ = CD′ = x(say) of the coil inside the magnetic field. If l is the length of the arm BC of the coil, then area of coil inside the magnetic field at any time t.

A = BCX CD′ = lx

Therefore magnetic flux linked with the coil at any time t.

f = BA = Blx

Suppose that the coil is pulled out of the magnetic field with velocity n. As the coil is pulled out magnetic flux linked with the coil changes. The time rate of change of magnetic flux linked with the coil is given by

If e is induced e.m.f. produced, then

$$\frac{d\phi}{dt} = \frac{-d}{dt}(Blx) = -Bl\frac{dx}{dt} = -Blv$$

e = -Bln

The negative sign shows that induced e.m.f. opposes to the coil being pulled out of magnetic field.

MUTUAL INDUCTION :

Consider two coil P and S are placed very close to each other. Coil P consists of battery and tapping key and coil S consists of galvanometer G. When the key of coil P is pressed then magnetic flux is building and induced e.m.f. produced in it opposes the flow of magnetic flux. Because coil P and coil S are very close to each other. So magnetic flux also changed in coil S and induced current is produced which opposes the direction of flow of magnetic lines of force in coil P.

The phenomenon according to which an opposing e.m.f. is produced in a coil as a result of change in current or magnetic flux linked with a neighboring coil is called mutual induction.

COEFFICIENT OF MUTUAL INDUCTION :

Suppose that current I is flowing through coil P and f be the magnetic flux linked with coil S

f a I

f = MI

M = Coefficient of mutual induction.

Let e be the induced e.m.f. in coil S.

$$e = \frac{d\phi}{dt} = \frac{-d}{dt}(MI) -$$ (-ive sign shows opposition of induced e.m.f.)

$$M = e / \frac{dI}{dt}$$

The mutual inductance of two coils is said to be one Henry, if a rate of change of current of 1 ampere per second in one coil induces an e.m.f. of 1 volt in neighboring coil.

SELF INDUCTION

Consider a coil connected to a battery and a tapping key. When key K is pressed magnetic lines of forces starts growing through it and induced e.m.f. is produced. Direction of induced e.m.f. is opposite to that of growth of current. On the other hand when key is released the current in the coil decreases and e.m.f. is produced in opposite direction. Thus during both growth and decay of current an opposite induced e.m.f. is produced. This e.m.f. is called back e.m.f.

The phenomenon according to which an opposing induced e.m.f. is produced in the coil as a result of change in current or magnetic flux linked with the coil is called self induction.

COEFFICIENT OF SELF INDUCTION :

Suppose when key is pressed, current I flows through the coil and magnetic flux f linked with the coil.

f a I

f = MI

L is called coefficient of self induction.

Let e be the induced e.m.f.

$$e = \frac{d\phi}{dt} = \frac{-d}{dt}(LI) = \frac{-LdI}{dt}$$

$$e = \frac{LdI}{dt}$$ (-ive sign shows opposing nature of induced e.m.f.)

$$M = e / \frac{dI}{dt}$$

Self inductance of a coil is said to be one Henry if a rate of change of current of 1 ampere per second induces an e.m.f. of one volt.

EDDY CURRENTS :

Eddy currents are the currents induced in a conductor, when placed in a changing magnetic field. They are also known as Focaults Currents.

Following experiment explain the origin of eddy currents. Introduce a soft iron core inside a solenoid and connect it to the source of alternating current. Place a metallic disk over soft iron core.

Explanation : When the circuit is switched on the current starts growing and hence magnetic flux linked with disk also increases. Thus disk is converted to small magnet. If soft iron's upper face acquires north polarity. Then metallic disk's lower surface acquires north polarity and due to repulsive force metallic disk placed over soft iron core is thrown up into the air.

APPLICATION OF EDDY CURRENTS :

DEAD BEAT GALVANOMETERS :

The oscillation of a moving coil galvanometer generally take a long time to die out. But by winding its coil on a metallic frame made of copper or aluminium the galvanometer can be made dead beat. It is because, due to production of eddy currents in a metallic frame. The coil of galvanometer comes to rest very soon.

SPEEDOMETER :

In speedometer, a small magnet is geared to the main shaft of the vehicle. The magnet is mounted in an aluminium cylinder with the help of hair springs. Due to rotation of magnet eddy currents are produced which led the drum to oppose the motion of relating magnet drum experience torque and gets deflected at certain angle.

ELECTRIC BRAKES :

A metallic drum is coupled to the wheels of train ; so that when train rotates drum also rotates. In order to stop the train magnetic field is applied to rotating drum. The eddy currents produced oppose the motion of drum. Since drum is connected to wheels of train, it comes to halt.

The following questions refer to the circuit shown in the figure, consisting of two flashlight batteries and two Nichrome wires of different lengths and different thicknesses as shown (corresponding roughly to your own thick and thin Nichrome wires). The thin wire is 55 cm long, and its diameter is 0.25 mm. The thick wire is 15 cm long, and its diameter is 0.32 mm.

(a) The emf of each flashlight battery is 1.5 volts. Determine the steady-state electric field inside each Nichrome wire. Remember that in the steady state you must satisfy both the current node rule and energy conservation. These two principles give you two equations for the two unknown fields.

E inside thin wire = V/m

E inside thick wire = V/m

(b) The electron mobility in room-temperature Nichrome is about 7×10 (m/s)/(N/C). How long (in minutes) does it take an electron to drift through the two Nichrome wires from location B to location A?

(c) On the other hand, about how long did it take to establish the steady state when the circuit was first assembled? Give an approximate, not precise numerical answer.

(d) There are about 9 10 mobile electrons per cubic meter in Nichrome. How many electrons cross the junction between the two wires every second?

electrons/s Suppose, I am trying to create a model with a very thin wire. The wire is supposed to be have a circular cross-section of 20 micron dia and be about 1 m in length (so, we are talking about a really thin, long wire).

I have trying adding cylinders with the right radius and length, and I have also tried drawing a circle with that radius and extruding. Both times I end up with something with a square cross-section.

I have a hunch that this might be a sampling error, but I do not know where to start. Any ideas ?I need to run a (sinusoidal) current through this very long wire and model the emf induced in a small inductor. I started with a quasistatic transient magnetic field model in 3D, but I am going through the documentation to find out how to setup a current. Do I have to setup the external circuit (apply a potential across the wire) or can just connect an imaginary harmonic current source in series with this somehow ?

The concept of electromotive force (emf) may be introduced in various ways in an undergraduate course of theoretical electromagnetism. The multitude of alternate expressions for the emf is often the source of confusion to the student. We summarize the main ideas, adopting a pedagogical logic that proceeds from the general to the specific. The emf of a "circuit" is first defined in the most general terms. The expressions for the emf of some familiar electrodynamical systems are then derived in a rather straightforward manner. A diversity of physical situations is thus unified within a common theoretical framework. 1. Introducin The difficulty in writing this article was not just due to the subject itself: we had to first overcome some almost irreconcilable differences in educational philosophy between an (opinionated) theoretical physicist and an (equally -if not more- opinionated) electrical engineer. At long last, a compromise was reached! This paper is the fruit of this "mutual understanding". Having taught intermediate-level electrodynamics courses for several years, we have come to realize that, in the minds of many of our students, the concept of elctromive force (emf) is something of a mystery. What is an emf, after all? Is it the voltage of an ideal battery in a DC circuit? Is it work per unit charge? Or is it, in a more sophisticated way, the line integral of the electric field along a closed path? And what if a magnetic rather than an electric field is present? Generally speaking, the problem with the emf lies in the diversity of situations where this concept applies, leading to a multitude of corresponding e 2 C. J. Papachristou, A. N. Magoulas 2. Thegnral definto m Consider a region of space in which an electromagnetic (e/m) field exists. In the most general sense, any closed path (or lop) within this region will be called a cirut (whether or not the whole or parts of consist of material objects such as wires, resistors, capacitors, batteries, or any other elements whose presence may contribute to the e/m field). We arbitly assign a positive direction of traversing the loop, and we consider an element dl of oriented in the positive direction. Imagine now a test charge located at the position of dl, and let be the force on at time : dl • + This force is exerted by the e/m field itself, as well as, possibly, by additional energy source (e.g., batteries) that can interact electrically with. The force punit charge at the position of dl at time, is = (1) Note that is independent of, since the force by the e/m field and/or the sources on is proportional to the charge. In particular, reversing the sign of will have no effect on (although it will change the direction of). We now define the elctromive frc (emf) of the circuit at time as the line integral of along, taken in the positve sense of : E = Å″ fdl +″ (2) Note that the sign of the emf is dependent upon our choice of the positive direction of circulation of : by changing this convention, the sign of E is reversed. We remark that, in the no-relativsc limit, the emf of a circuit is the same for all inertial observers since at this limt the force is invariant under a change of frame of reference. In the following sections we apply the defining equation (2) to a number of specific electrodynamic situations that are certainly familiar to the student. Electromotive force: A guide for the perplexed 3 3. Motinal emfinth presncof atimgne ic feld Conside a circuit consisting of a closed wire. The wire is moving inside a staic magnetic field Br ( ).

Let õ be the velocity of the element dl of relative to our inertial frame of reference. A charge (say, a free electron) at the location of dl executes a composite motion, due to the motion of the loop itself relative to our frame, as well as the motion of along. The total velocity of relative to us is õ õ õ to = + 2, where õ2 is the velocity of in a direction parallel to dl. The force from the magnetic field on is ( ) ( ) ( ) ( ) ( ) to FqBqBqB fBB õ õ õ õ õ = × = × + × Ò!2 = = × + ×2 By (2), then, the emf of the circuit is E ( ) ( ) CCC = Å″ = ×Å″ + ×Å″ fdlBdlBdl õõ2 +″ +″ +″ But, since õ2 is parallel to dl, we have that ( ) 0 õ2 × Å″ = Bdl. Thus, finally, E = × Å″ ( ) õ Bdl +″ (3) Note that the wire ned otmain fixed shap,ize orienta during its motion! Note also that the velocity õ may vary around the circuit. By using (3), it can be proven (see Appendix) that E dt Ö = ″ (4) where Ö = Å″ Bda +″ is the magnetic flux through the wire at time. Note carefully that (4) does not express any novel physical law: it is simply a direct consequence of the definition of the emf ! 4. Emfdueto aim-vryng maeticfld Consider now a closed wire that is at rest inside a time-varying magnetic field Brt (,).

As experiments show, as soon as starts changing, a current begins to flow in the wire. This looks impressive, given that the free charges in the (stationary) wire were initially at rest. And, as everybody knows, a magnetic field exerts forces on moving charges only! It is also observed experimentally that, if the magnetic field stops varying in time, the current in the wire disappears. The only field that can put an initially stationary charge in motion and keep this charge moving is an elctri field. 4 C. J. Papachristou, A. N. Magoulas We are thus compelled to conclude that a time-varying maeticfld snecarily acompnied by an elctri field. (It is often said that "a changing magnetic field induces an electric field". This is somewhat misleading since it gives the impression that the "source" of an electric field could be a magnetic field. Let us keep in mind, however, that the true sources of any e/m field are the electric charges and the electric currents!) So, let Ert (,) be the electric field accompanying the time-varying magnetic field. Consider again a charge at the position of the element dl of the wire. Given that the wire is now at rest (relative to our inertial frame), the velocity of will be due to the motion of the charge along the wire only, i.e., in a direction parallel to dl : õ õ to = 2 (since õ = 0). The force on by the e/m field is [( )] [( )] ( ) to FqEBqEB fEB õ õ õ = + × = + × Ò!2 = = + ×2 The emf of the circuit is now E ( ) CCC = Å" = Å" + ×Å" fdlEdlBdlõ2 +" +" +" But, as explained earlier, ( ) 0 õ2 × Å" = Bdl. Thus, finally, E = Å"Å dl +" (5) Equation (4) is still valid. This time, however, it is not merely a mathematical consequence of the definition of the emf ; rather, it is a true physical law deduced from experiment! Let us examine it in some detail. In a region of space where a time-varying e/m field (,) EB exists, consider an arbitrary open surface bounded by the closed curve : da da dl (The relativ direction of dl and the surface element da, normal to, is determined according to the familiar right-hand rule.) The loop is assumed stationary relative to the inertial observer; hence the emf along at time is given by (5).

(Note that the signs of E and Ö depend on the chosen positive direction of.) Since the field is solenida, the value of Ö for a given is independent of the choice of the surface. That is, the same magnetic flux will go through open surface bounded by the closed curve. According to the Fardy-Hen law, E dt Ö = " (6) or explicitly, CS EdlBda dt Å" = " Å"++" (7) (The negative sign on the right-hand sides of (6) and (7) expresses Lenz'slaw.) Equation (7) can be re-expressed in differential form by using Stokes' theorem, ( ) CS EdlEda Å" = "×Å"++" and by taking into account that the surface may be arbitrarily chosen. The result is " "× = " " (8` We note that if " " `" Bt / 0, then necessarily `" 0. Hence, as already mentioned, a time-varying magnetic field is always accompanied by an electric field. If, however, is staic (" " = Bt / 0), then is irotanl : "× = Ô!Å" = EEdl 0 0 +", which allows for the possibility that = 0. Corlay: The emf around a fixed loop inside a staic e/m field (ErBr ( ), ( )) is E = 0 (the student should explain this). 5. Emf of a circuit containing a battery and a resistor Consider a circuit consisting of an ideal battery (i.e., one with no internal resistance) connected to an external resistor. As shown below, the emf of the circuit in thedrction of the curent is equal to the voltage of the battery. Moreover, the emf in this case represents the work peunitcharge done by th 6 C. J. Papachristou, A. N. Magoulas i i + _ 0 We recall that, in general, the emf of a circuit at time is equal to the integral E = Å" fdl +" where fFq = / is the force per unit charge at the location of the element dl of the circuit, at time. In essence, we assume that in every element dl we have placed a test charge (this could be, e.g., a free electron of the conducting part of the circuit). The force on each is then measured simultaneosy for all charges at time. Since here we are dealing with a staic (time-independent) situation, however, we can treat the problem somewhat differently: The measurements of the forces on the charges need not be made at the same instant, given that nothing changes with time, anyway. So, instead of placing several charges around the circuit and measuring the forces on each of them at a particular instant, we imagine a singlechar making a complete tour around the loop. We may assume, e.g., that the charge is one of the (convetialy positve) free electrons taking part in the constant current É flowing in the circuit. We then measure the force on at each point of. We thus assume that is a positve charge moving in thedrcion fthecurn É. We also assume that the direction of circulation of is the same as the directon of the curent (counterclockwise in the figure). During its motion, is subject to two forces: (1) the force 0 by the source (battery) that carries from the negative pole to the positive pole b throug thesourc, and (2) the electrostatic force FqE = due to the electrostatic field at each point of the circuit (both inside and outside the source). The total force on is 0 0 0 0 FF FFFFqEfEfE qq = + = + Ò! = = + a″ + Then, E 0 0 CCCC = Å″ = Å″ + Å″ = Å″ fdlfdlEdlfdl ++++(9) since Edl Å″ = 0 +″ for an electrostatic field. However, the action of the source on is limited to the region between the poles of the battery, that is, the section of the circuit from to. Hence, 0 = 0 outside the source, so that (9) reduces to Electromotive force:

A guide for the perplexed 7 E 0 = Å″ fdl +″ (10) Now, since the current É is constant, the charge moves at constant speed along the circuit. This means that the toal force on in the direction of the path is zero. In the interior of the resistor, the electrostatic force FqE = is counterbalanced by the force on due to the collisions of the charge with the positive ions of the metal (this latter force does not contribute to the emf and is not counted in its evaluation!). In the interior of the (ideal) battery, however, where there is no resistance, the electrostatic force must be counterbalanced by the opsing force 0 exerted by the source. Thus, in the section of the circuit between and, 0 0 0 FFFffEfE = + = Ò! = = + = Ò! = ″ 0 0 Equation (10) then takes the final form, E ba = ″ Å″ = ″ = EdlVVV +″ (11) where and are the electrostatic potentials at and, respectively. This is, of course, what every student knows from elementary e/m courses.

The work done by the source on upon transferring the charge from to is 0 0 bb aa WFdlqfdlq = Å″ = Å″ = ++″ E (12) [where we have used (10)]. So, the work fthesourc penitharge is /q= E. This work is converted into heat in the resistor, so that the source must again supply energy in order to carry the charges once more from to b. This is something like the torture of Sisyphus in Greek mythology! 6. Emf and Ohm's law Consider a closed wire inside an e/m field. The circuit may contain sources (e.g., a battery) and may also be in motion relative to our inertial frame of reference. Let be a test charge at the location of the element dl of, and let be the total force on (due to the e/m field and/or the sources) at time. (As mentioned in Sec.2, this force is, classically, a frame-independent quantity.) The force per unit charge at the location of dl at time, then, is fFq = /. According to our general definition, the emf of the circuit at time is E = Å″ fdl +″ (13) 8 C. J. Papachristou, A. N. Magoulas Now, if ó is the condutivy of the wire, then, by Ohm's law in its general form (see, e.g., p. 285 of) we have: Jf = ó (14) where is the volume curent density at the location of dl at time. (Note that the more common expression JE =ó, found in most textbooks, is a special case of the above formula. Note also that is measured relativ ohewir, thus is the same for all inertial observers.) By combining (13) and (14) we get: E 1 Jdl ó = Å"+" (15) Taking into account that is in the direction of dl at each point of, we write: JdlJdldl Å" = =

7. Concluding remarks In concluding this article, let us highlight a few points of importance: 1. The emf was defined as a line integral of force per unit charge around a loop (or "circuit") in an e/m field. The loop may or may not consist of a real conducting wire, and it may contain sources such as batteries. 2. In the classical (non-relativistic) limit, the emf is independent of the inertial frame of reference with respect to which it is measured. 10 C. J. Papachristou, A. N. Magoulas 3. In the case of purely motinal emf, Faraday's "law" (4) is in essence a mere consequence of the definition of the emf. On the contrary, when a time-dependent magnetic field is present, the similar-looking equation (6) is a true physical law (the Faraday-Henry law). 4. In a DC circuit with a battery, the emf in the direction of the current equals the voltage of the battery and represents work per unit charge done by the source. 5. If the loop describing the circuit represents a conducting wire of finite resistance, Ohm's law can be expressed in terms of the emf by equation (16).

7

Two phenomena and Einstein's View

Consider Faraday's flux law for the EMF generated in a conductor loop: where åå is the EMF, and ÕÕ is the magnetic flux through the loop.

There are two possible causes for the flux to variate over time: variations in the magnetic field ("transformer EMF") and variations in the area enclosed by the loop ("motional EMF").

Feynman has noted that this is a unique case where a single rule is explained by two different phenomena:

We know of no other place in physics where such a simple and accurate general principle requires for its real understanding an analysis in terms of two different phenomena. Usually such a beautiful generalization is found to stem from a single deep underlying principle. Nevertheless, in this case there does not appear to be any such profound implication.

RICHARD P. FEYNMAN, THE FEYNMAN LECTURES ON PHYSICS (VOLUME II, 17-2).

Is it really true that there is no way to view these two phenomena as one?

For example, it says in this Wikipedia article that this apparent dichotomy was part of what led Einstein to develop special relativity.

Does special relativity give us a unifying principle to derive Faraday's law from?

he two phenomena Feynman referred to are the qv× ×B× qv′!×B′! part of the force acting on an electric charge carrier; and the ″×E× =┐″tB× ″×E′!=┐″tB′! dynamical Maxwell's equation. In the most general situation when both the magnetic field and the shape of the wire is changing, we have to use and add both terms. That's true in each reference frame because for a complicated space-dependent, time-dependent geometry of the wires and fields, there won't be any inertial frame in which one of the phenomena would completely vanish.

I believe that it's more likely (but not certain) Feynman only meant this thing that there's no way to eliminate or "explain" one of the terms in the general situation.

On the other hand, the fact that both terms have the same origin is a consequence of special relativity and it was indeed one of the motivations that led Einstein to his new picture of spacetime. It seems somewhat plausible to me that Feynman was ignorant about this history.

The full theory of electrodynamics is nicely Lorentz-covariant and it implies both terms. However, these terms aren't not really the same. The Lorentz force comes from the integral qAìdxìqAìdxì over the world lines of charged particles while Maxwell's equations arise from the " FìíFìí" FìíFìí Maxwell Lagrangian. So Feynman would also be right if he said that the two terms can't be transformed to each other by any symmetry transformation.

Still, one can make physical arguments that do involve such transformations and imply that the two phenomena are inseparable. For example, if we assume the Maxwell equation, it follows from the Lorentz symmetry that E× E'! must transform as the remaining 3 components of the antisymmetric tensor whose purely spatial components give B× B'!. But then it follows that the force acting on a charged particle, qE× qE'!, must also be extended by the remaining term qv× ×B× qv'!×B'! for the theory to be Lorentz-invariant. Or one can run the argument backwards. Still, we are dealing with transformations of two different terms in the action that just happen to have a "unified, simply describable" impact on the EMF in wires with magnetic flux.

The simplification and unity only occurs if we assume that the two different kinds of phenomena are in the action to start with but they deal with the same fields which respect the same Lorentz symmetry; and if we study situations that are "understood" or "simplified" on both sides in which some effects, e.g. the magnetic ones, are absent.

Let me say it differently: if the area enclosed by a wire goes to zero, it's an objective thing that is clearly independent of the reference frame. So one shouldn't expect the shrinking of the area is just a matter of inertial systems; it's a frame-independent fact. On the other hand, it's not shocking that the EMF ultimately only depends on one thing, the change of the flux, which has a simple form although it may have different origin.

I would conclude that Feynman was more right than wrong. Lorentz symmetry operates in both phenomena and it's the same one which is a constraint on the most general theory; however, the fact that both possible sources of the changing flux influence the EMF in the same way is a sort of "coincidence", at least if we use the conventional variables to describe the electromagnetic phenomena.

- Parts of which I'm still struggling to understand. Two questions: In the third paragraph from the end, I didn't understand what situations you refer to, in which the simplification and unity can supposedly occur. Or better yet, in what situations do they not occur? Also, I understand the fact that both phenomena influence the EMF in the same way is a "coincidence" if we use the conventional variables, but is it still a coincidence in covariant variables? Is there a covariant formulation of the flux law? – Joe Feb 1 '13 at 20:46
- Dear Joe, by the simplification, I meant the fact that the EMF only depends on the change of flux and not on some more complicated function of changes of the area, the magnetic field, or some combination of them. When one considers the induction in the case of uniform magnetic fields only etc., this simplification for this formula for EMF always occurs. But as Feynman correctly suggests, this simplification does not imply that the terms that combine to the full answer are related by symmetries or "unified" in the conventional sense. – Luboš Motl Feb 3 '13 at 7:44
- Otherwise, I was using (Lorentz) covariant variables all the time and the terms don't and didn't unify into "a term of the same origin" so yes, covariance isn't enough to explicitly explain the simplification. It was the bulk and key point of my answer, I feel that I am already answering the same question for the 4th time. So no, Feynman wasn't wrong in his statement. Maybe you don't like the answer because you were predecided that the right answer should be very different? – Luboš Motl Feb 3 '13 at 7:46
- In fact I did, but I was wrong. Sorry for being slow to understand this, and thank you for all the explanations. – Joe Feb 3 '13 at 7:52
- Luboš and @Joe : It seems to me that the Maxwell-Ampere law also raises the "two different phenomena" issue. This way of looking at it turns out to be useful

Electromagnetism and Relativity We've seen that Maxwell's equations have wave solutions which travel at the speed of light. But there's another place in physics where the speed of light plays a prominent role: the theory of special relativity. How does electromagnetism fit with special relativity? Historically, the Maxwell equations were discovered before the theory of special relativity. It was thought that the light waves we derived above must be oscillations of some substance which fills all of space. This was dubbed the aether. The idea was that Maxwell's equations only hold in the frame in which the aether is at rest; light should then travel at speed c relative to the aether. We now know that the concept of the aether is unnecessary baggage. Instead, Maxwell's equations hold in all inertial frames and are the first equations of physics which are consistent with the laws of special relativity. Ultimately, it was by studying the Maxwell equations that Lorentz was able to determine the form of the Lorentz transformations which subsequently laid the foundation for Einstein's vision of space and time. Our goal in this section is to view electromagnetism through the lens of relativity. We will find that observers in diµ!erent frames will disagree on what they call electric fields and what they call magnetic fields. They will observe diµ!erent charge densities and diµ!erent currents. But all will agree that these quantities are related by the same Maxwell equations. Moreover, there is a pay-oµ! to this. It's only when we formulate the Maxwell equations in a way which is manifestly consistent with relativity that we see their true beauty. The slightly cumbersome vector calculus equations that we've been playing with throughout these lectures will be replaced by a much more elegant and simple-looking set of equations. 5.1 A Review of Special Relativity We start with a very quick review of the relevant concepts of special relativity. (For more details see the lecture notes on Dynamics and Relativity). The basic postulate of relativity is that the laws of physics are the same in all inertial reference frames. The guts of the theory tell us how things look to observers who are moving relative to each other. The first observer sits in an inertial frame S with spacetime coordinates (ct, x, y, z) the second observer sits in an inertial frame S0 with spacetime coordinates (ct0, x0, y0, z0). – 94 – If we take S0 to be moving with speed v in the x-direction relative to S then the coordinate systems are related by the Lorentz boost x0 = ã! x v c ct'# and ct0 = ã! ct v c x '# (5.1) while y0 = y and z0 = z. Here c is the speed of light which has the value, c = 299792458 ms1 Meanwhile is the ubiquitous factor = s 1 1 v2/c2 The Lorentz transformation (5.1) encodes within it all of the fun ideas of time dilation and length contraction that we saw in our first course on relativity. 5.1.1 Four-Vectors It's extremely useful to package these spacetime coordinates in 4-vectors, with indices running from µ = 0 to µ = 3 Xµ = (ct, x, y, z) µ = 0, 1, 2, 3 Note that the index is a superscript rather than subscript. This will be important shortly.

A general Lorentz transformation is a linear map from X to X0 of the form (X0) µ = ä!µ +#X+# Here ä! is a 4 å! 4 matrix which obeys the matrix equation ä!T '#ä! = '#, ä!â! µ'#â!ä!+# = '#µ+# (5.2) with '#µ+# the Minkowski metric '#µ+# = diag(+1, 1, 1, 1) The solutions to (5.2) fall into two classes. The first class is simply rotations. Given a 3 å! 3 rotation matrix R obeying RTR = 1, we can construct a Lorentz transformation ä! obeying (5.2) by embedding R in the spatial part, ä!µ +# = 0 BBBB@ 1 000 0 0 R 0 1 CCCCA (5.3) These transformations describe how to relate the coordinates of two observers who are rotated with respect to each o The other class of solutions to (5.2) are the Lorentz boosts. These are the transformations appropriate for observers moving relative to each other. The Lorentz transformation (5.1) is equivalent to ä!µ +# = 0 BBBB@ v/c 0 0 v/c 0 0 0 0 1 0 0 0 0 1 1 CCCCA (5.4) There are similar solutions associated to boosts along the y and z axes. The beauty of 4-vectors is that it's extremely easy to write down invariant quantities. These are things which all observers, no matter which their reference frame, can agree on. To construct these we take the inner product of two 4-vectors. The trick is that this inner product uses the Minkowski metric and so comes with some minus signs. For example, the square of the distance from the origin to some point in spacetime labelled by X is X · X = Xµ'#µ+#X+# = c2 t 2 x2 y2 z2 which is the invariant interval. Similarly, if we're given two four-vectors X and Y then the inner product X · Y = Xµ'#µ+#Y +# is also a Lorentz invariant. 5.1.2 Proper Time The key to building relativistic theories of Nature is to find the variables that have nice properties under Lorentz transformations.

The 4-vectors X, labelling spacetime points, are a good start. But we need more. Here we review how the other kinematical variables of velocity, momentum and acceleration fit into 4-vectors. Suppose that, in some frame, the particle traces out a worldline. The clever trick is to find a way to parameterise this path in a way that all observers agree upon. The natural choice is the proper time '#, the duration of time experienced by the particle itself. If you're sitting in some frame, watching some particle move with an old-fashioned Newtonian 3-velocity u(t), then it's simple to show that the relationship between your time t and the proper time of the particle '# is given by dt d'# = (The proper time allows us to define the 4-velocity and the 4-momentum. Suppose that the particle traces out a path X('#) in some frame. Then the 4-velocity is U = dX d'# = c u ! Similarly, the 4-momentum is P = mU where m is the rest mass of the particle. We write P = E/c p ! (5.5) where E = mc2 is the energy of the particle and p = mu is the 3-momentum in special relativity. The importance of U and P is that they too are 4-vectors. Because all observers agree on '#, the transformation law of U and P are inherited from X. This means that under a Lorentz transformation, they too change as U !ä!U and P !ä!P. And it means that inner products of U and P are guaranteed to be Lorentz invariant. 5.1.3 Indices Up, Indices Down Before we move on, we do need to introduce one extra notational novelty.

The minus signs in the Minkowski metric '# means that it's useful to introduce a slight twist to the usual summation convention of repeated indices. For all the 4-vectors that we introduced above, we always place the spacetime index µ = 0, 1, 2, 3 as a superscript (i.e. up) rather than a subscript. Xµ = ct x ! This is because the same object with an index down, Xµ, will mean something subtly diµ!erent. We define Xµ = ct x ! With this convention, the Minkowski inner product can be written using the usual convention of summing over repeated indices as XµXµ = c2 t 2 x · x In contrast, XµXµ = c2t 2 + x2 is a dumb thing to write in the context of special relativity since it looks very diµ!erent to observers in diµ!erent inertial frames. In fact, we will shortly declare it illegal to write things like XµXµ. – There is a natural way to think of Xµ in terms of Xµ using the Minkowski metric '#µ+# = diag(+1, 1, 1, 1). The following equation is trivially true: Xµ = '#µ+#X+#This means that we can think of the Minkowski metric as allowing us to lower indices. To raise indices back up, we need the inverse of '#µ+# which, fortunately, is the same matrix: '#µ+# = diag(+1, 1, 1, 1) which means we have '#µâ!'#â!+# = µ +# and we can write X+# = '#+#µXµ From now on, we're going to retain this distinction between all upper and lower indices. All the four-vectors that we've met so far have upper indices. But all can be lowered in the same way. For example, we have Uµ = c u ! (5.6) This trick of distinguishing between indices up and indices down provides a simple formalism to ensure that all objects have nice transformation properties under the Lorentz group. We insist that, just as in the usual summation convention, repeated indices only ever appear in pairs. But now we further insist that pairs always appear with one index up and the other down.

The result will be an object which is invariant under Lorentz transformations. 5.1.4 Vectors, Covectors and Tensors In future courses, you will learn that there is somewhat deeper mathematics lying behind distinguishing Xµ and Xµ: formally, these objects live in diµ!erent spaces (sometimes called dual spaces). We'll continue to refer to Xµ as vectors, but to distinguish them, we'll call Xµ covectors. (In slightly fancier language, the components of the vector Xµ are sometimes said to be contravariant while the components of the covector Xµ are said to be covariant). For now, the primary diµ!erence between a vector and covector is how they transform under rotations and boosts. We know that, under a Lorentz transformation, any 4-vector changes as Xµ ! X0 µ = ä!µ +#X From this, we see that a covector should transform as Xµ ! X0 µ = '#µâ!X0 â! = '#µâ!ä!â! X = '#µâ!ä!â!'#+#X+# Using our rule for raising and lowering indices, now applied to the Lorentz transformation ä!, we can also write this as Xµ ! ä!+# µ X+# where our notation is now getting dangerously subtle: you have to stare to see whether the upper or lower index on the Lorentz transformation comes first.

There is a sense in which ä!+# µ can be thought of a the components of the inverse matrix ä!1. To see this, we go back to the definition of the Lorentz transformation (5.2), and start to use our new rules for raising and lowering indices ä!â! µ'#â!ä!+# = '#µ+#) ä!â! µä!â!+# = '#µ+#) ä!â! µä!â! = µ) ä!â!ä!â! µ = µ In the last line above, we've simply reversed the order of the two terms on the left. (When written in index notation, these are just the entries of the matrix so there's no problem with commuting them). Now we compare this to the formula for the inverse of a matrix, (ä!1)â!ä!â! µ = µ) (ä!1) â! = ä!â! (5.7) Note that you need to be careful where you place the indices in equations like this. The result (5.7) is analogous to the statement that the inverse of a rotation matrix is the transpose matrix. For general Lorentz transformations, we learn that the inverse is sort of the transpose where "sort of" means that there are minus signs from raising and lowering. The placement of indices in (5.7) tells us where those minus signs go. The upshot of (5.7) is that if we want to abandon index notation all together then vectors transform as X !ä!X while covectors – which, for the purpose of this sentence, we'll call X˜ – transform as X˜ !ä!1X˜. However, in what follows, we have no intention of abandoning index notation. Instead, we will embrace it. It will be our friend and our guide in showing that the Maxwell equations are consistent with specia A particularly useful example of a covector is the four-derivative. This is the relativistic generalisation of r, defined by @µ = @ @Xµ = –'1 c @ @t, r Æ% Notice that the superscript on the spacetime 4-vector Xµ has migrated to a subscript on the derivative @µ. For this to make notational sense, we should check that @µ does indeed transform as covector. This is a simple application of the chain rule. Under a Lorentz transformation, Xµ ! X0 µ = ä!µ +#X+#, so we have @µ = @ @Xµ ! @ @X0 µ = @X+# @X0 µ @ @X+# = (ä!1)+# µ@+# = ä!+# µ @+# which is indeed the transformation of a co-vector. Tensors Vectors and covectors are the simplest examples of objects which have nice transformation properties under the Lorentz group. But there are many more examples.

The most general object can have a bunch of upper indices and a bunch of lower indices, T µ1...µn +#1...+#m. These objects are also called tensors of type (n, m). In order to qualify as a tensor, they must transform under a Lorentz transformation as T0 µ1...µn +#1...+#m = ä!µ1 â!1... ä!µn â!n ä! 1 +#1... ä! m +#m Tâ!1...â!n 1...m (5.8) You can always use the Minkowski metric to raise and lower indices on tensors, changing the type of tensor but keeping the total number of indices n + m fixed. Tensors of this kind are the building blocks of all our theories. This is because if you build equations only out of tensors which transform in this manner then, as long as the µ, +#,... indices match up on both sides of the equation, you're guaranteed to have an equation that looks the same in all inertial frames. Such equations are said to be covariant. You'll see more of this kind of thing in courses on General Relativity and Diµ!erential Geometry. In some sense, this index notation is too good. Remember all those wonderful things that you first learned about in special relativity: time dilation and length contraction and twins and spaceships so on. You'll never have to worry about those again. From now on, you can guarantee that you're working with a theory consistent with relativity by ensuring two simple things

- That you only deal with tensors.
- That the indices match up on both sides of the equation. It's sad, but true. It's all part of growing up and not having fun anymore. – 5.2 Conserved Currents We started these lectures by discussing the charge density â!(x, t), the current density J(x, t) and their relation through the continuity equation, @â! @t + r · J = 0 which tells us that charge is locally conserved. The continuity equation is already fully consistent with relativity. To see this, we first need to appreciate that the charge and current densities sit nicely together in a 4-vector, Jµ = â!c J ! Of course, placing objects in a four-vector has consequence: it tells us how these objects look to diµ!erent observers. Let's quickly convince ourselves that it makes sense that charge density and current do indeed transform in this way. We can start by considering a situation where there are only static charges with density â!0 and no current. So Jµ = (â!0, 0). Now, in a frame that is boosted by velocity v, the current will appear as J0 µ = ä!µ +#J+# with the Lorentz transformation given by (5.4). The new charge density and current are then â!0 = â!0, J0 = â!v The first of these equations tells us that diµ!erent observers see diµ!erent charge densities. This is because of Lorentz contraction: charge density means charge per unit volume. And the volume gets squeezed because lengths parallel to the motion undergo Lorentz contraction.

That's the reason for the factor of in the observed charge density. Meanwhile, the second of these equations is just the relativistic extension of the formula J = â!v that we first saw in the introduction. (The extra minus sign is because v here denotes the velocity of the boosted observer; the charge is therefore moving with relative velocity v). In our new, relativistic, notation, the continuity equation takes the particularly simple form @µJµ = 0 (5.9) This equation is Lorentz invariant. This follows simply because the indices are contracted in the right way: one up, and one down. – 5.2.1 Magnetism and Relativity We've learned something unsurprising: boosted charge gives rise to a current. But, combined with our previous knowledge, this tells us something new and important: boosted electric fields must give rise to magnetic fields. The rest of this chapter will be devoted to understanding the details of how this happens. But first, we're going to look at a simple example where we can re-derive the magnetic force purely from the Coulomb force and a dose of Lorentz contraction. To start, consider a bunch of positive charges A v v u Figure 45: +q moving along a line with speed +v and a bunch of negative charges q moving in the opposite direction with speed v as shown in the figure. If there is equal density, n, of positive and negative charges then the charge density vanishes while the current is I = 2nAqv where A is the cross-sectional area of the wire. Now consider a test particle, also carrying charge q, which is moving parallel to the wire with some speed u. It doesn't feel any electric force because the wire is neutral, but we know it experiences a magnetic force. Here we will show how to find an expression for this force without ever invoking the phenomenon of magnetism. The trick is to move to the rest frame of the test particle. This means we have to boost by speed u.

The usual addition formula tells us that the velocities of the positive and negative charges now diµ!er, given by v± = v %# u 1 %# uv/c2 But with the boost comes a Lorentz contraction which means that the charge density changes. Moreover, because the velocities of positive and negative charges are now diµ!erent, this will mean that, viewed from the rest frame of our particle, the wire is no longer neutral. Let's see how this works. First, we'll introduce n0, the density of charges when the particles in the wire are at rest. Then the charge density in the original frame is â! = qn = (v)qn0 In this frame the wire is neutral because the positive and negative charges travel at the same speed, albeit in opposite directions. However, in our new frame, the charge – 10 densities are â!± = qn± = q(v±)n0 = ã! 1 %# uv c2 '# (u)(v) qn0 where you've got to do a little bit of algebra to get to the last result. Since v > v+, we have n > n+ and the wire carries negative charge. The overall net charge density in the new frame is â!0 = qn0 = q(n+ n) = 2uv c2 (u) qn But we know that a line of electric charge creates an electric field; we calculated it in (2.6); it is E(r) = 2uv c2 (u) qnA 2á!'0r ˆr where r is the radial direction away from the wire. This means that, in its rest frame, the particle experiences a force F0 = u(u) nAq2v á!'0c2r where the minus sign tells us that the force is towards the wire for u > 0. But if there's a force in one frame, there must also be a force in another. Transforming back to where we came from, we conclude that even when the wire is neutral there has to be a force F = F0 (u) = u nq2Av á!'0c2r = uq µ0I 2á!r (5.10) But this precisely agrees with the Lorentz force law, with the magnetic field given by the expression (3.5) that we computed for a straight wire. Notice that if u > 0 then the test particle – which has charge q – is moving in the same direction as the particles in the wire which have charge q and the force is attractive. If u < 0 then it moves in the opposite direction and the force is repulsive.

This analysis provides an explicit demonstration of how an electric force in one frame of reference is interpreted as a magnetic force in another. There's also something rather surprising about the result. We're used to thinking of length contraction as an exotic result which is only important when we approach the speed of light. Yet the electrons in a wire crawl along. They take around an hour to travel a meter! Nonetheless, we can easily detect the magnetic force between two wires which, as we've seen above, can be directly attributed to the length contraction in the electron d The discussion above needs a minor alteration for actual wires. In the rest frame of the wire the positive charges – which are ions, atoms stripped of some of their electrons – are stationary while the electrons move. Following the explanation above, you might think that there is an imbalance of charge density already in this frame. But that's not correct. The current is due to some battery feeding electrons into the wire and taking them out the other end. And this is done in such a way that the wire is neutral in the rest frame, with the electron density exactly compensating the ion density.

In contrast, if we moved to a frame in which the ions and electrons had equal and opposite speeds, the wire would appear charged. Although the sta ting point is slightly diµ!erent, the end result remains. 5.3 Gauge Potentials and the Electromagnetic Tensor Under Lorentz transformations, electric and magnetic fields will transform into each other. In this section, we want to understand more precisely how this happens. At first sight, it looks as if it's going to be tricky. So far the objects which have nice transformation properties under Lorentz transformations are 4-vectors. But here we've got two 3-vectors, E and B. How do we make those transform into each other? 5.3.1 Gauge Invariance and Relativity To get an idea for how this happens, we first turn to some objects that we met previously: the scalar and vector potentials and A. Recall that we introduced these to solve some of the equations of electrostatics and magnetostatics, r å! E = 0) E = r r · B = 0) B = r å! A However, in general these expressions can't be correct. We know that when B and E change with time, the two source-free Maxwell equations are r å! E + @B @t = 0 and r · B = 0 Nonetheless, it's still possible to use the scalar and vector potentials to solve both of these equations.

The solutions are E = r @A @t and B = r å! A where now = (x, t) and A = A(x, t). Just as we saw before, there is no unique choice of and A. We can always shift A ! A + r and B remains unchanged. However, now this requires a compensating shift of. ! @ @t and A ! A + r (5.11) with = (x, t). These are gauge transformations. They reproduce our earlier gauge transformation for A, while also encompassing constant shifts in. How does this help with our attempt to reformulate electromagnetism in a way compatible with special relativity? Well, now we have a scalar, and a 3-vector: these are ripe to place in a 4-vector. We define Aμ = /c A ! Or, equivalently, Aμ = (/c, A). In this language, the gauge transformations (5.11) take a particularly nice form, Aμ ! Aμ @μ (5.12) where is any function of space and time 5.3.2 The Electromagnetic Tensor We now have all the ingredients necessary to determine how the electric and magnetic fields transform. From the 4-derivative @μ = (@/ @(ct), r) and the 4-vector Aμ = (/c, A), we can form the anti-symmetric tensor Fμ+# = @μA+# @+#Aμ This is constructed to be invariant under gauge transformations (5.12). We have Fμ+# ! Fμ+# + @μ@+# @+#@μ = Fμ+# This already suggests that the components involve the E and B fields. To check that this is indeed the case, we can do a few small computations, F01 = 1 c @(Ax) @t @(/ c) @x = Ex c and F12 = @(Ay) Similar computations for all other entries give us a matrix of electric and magnetic fields, Fμ+# = 0 BBBB@ 0 Ex/c Ey/c Ez/c Ex/c 0 Bz By Ey/c Bz 0 Bx Ez/c By Bx 0 1 CCCCA (5.13) This, then, is the answer to our original question. You can make a Lorentz covariant object consisting of two 3-vectors by arranging them in an anti-symmetric tensor. Fμ+# is called the electromagnetic tensor.

Equivalently, we can raise both indices using the Minkowski metric to get F μ+# = '#μâ!'#+#Fâ! = 0 BBBB@ 0 Ex/c Ey/c Ez/c Ex/c 0 Bz By Ey/c Bz 0 Bx Ez/c By Bx 0 1 CCCCA Both Fμ+# and F μ+# are tensors. They are tensors because they're constructed out of objects, Aμ, @μ and '#μ+#, which themselves transform nicely under the Lorentz group. This means that Fμ+# must transform as F0 μ+# = ä!μ â!ä!+# Fâ! (5.14) Alternatively, if you want to get rid of the indices, this reads F0 = ä!Fä!T. The observer in a new frame sees electric and magnetic fields E0 and B0 that diμ!er from the original observer. The two are related by (5.14). Let's look at what this means in a couple of illustrative examples. Rotations To compute the transformation (5.14), it's probably simplest to just do the sums that are implicit in the repeated â! and labels. Alternatively, if you want to revert to matrix multiplication then this is the same as F0 = ä!Fä!T. Either way, we get the same result. For a rotation, the 3 å! 3 matrix R is embedded in the lower-right hand block of ä! as shown in (5.3). A quick calculation shows that the transformation of the electric and magnetic fields in (5.14) is as expected, E0 = R E and Boosts Things are more interesting for boosts.

Let's consider a boost v in the x-direction, with ä! given by (5.4). Again, you need to do a few short calculations. For example, we have E0 x c = F0 01 = ä!0 â!ä!1 Fâ! = ä!0 0ä!1 1F01 + ä!0 1ä!1 0F10 = 2v2 c2 Ex c 2Ex c = Ex c and E0 y c = F0 02 = ä!0 â!ä!2 Fâ! = ä!0 0ä!2 2F02 + ä!0 1ä!2 2F12 = Ey c + v c Bz = c (Ey vBz) and B0 z = F0 12 = ä!1 â!ä!2 Fâ! = ä!1 0ä!2 2F02 + ä!1 1ä!2 2F12 = v c2 Ey Bz = (Bz vEy/c2) The final result for the transformation of the electric field after a boost in the x-direction is E0 x = Ex E0 y = (Ey vBz) (5.15) E0 z = (Ez + vBy) and, for the magnetic field, B0 x = Bx B0 y = ã! By + v c2Ez '# (5.16) B0 z = ã! Bz v c2Ey '# As we anticipated above, what appears to be a magnetic field to one observer looks like an electric field to an Note that in the limit v '# c, we have E0 = E + v å! B and B0 = B. This can be thought of as the Galilean boost of electric and magnetic fields. We recognise E+vå!B as the combination that appears in the Lorentz force law. We'll return to this force in Section 5.5 where we'll see how it's compatible with special relativity. 5.3.3 An Example: A Boosted Line Charge In Section 2.1.3, we computed the electric field due to a line with uniform charge density '# per unit length. If we take the line to lie along the x-axis, we have (2.6) E = '# 2á!'0(y2 + z2) 0 y z ! (5.17) Meanwhile, the magnetic field vanishes for static electric charges: B = 0. Let's see what this looks like from the perspective of an observer moving with speed v in the x-direction, parallel to the wire. In the moving frame the electric and magnetic fields are given by (5.15) and (5,16). These read E0 = '# 2á!'0(y2 + z2) 0 y z ! = '# 2á!'0(y0 2 + z0 2) 0 y0 z0 ! B0 = '#v 2á!'0c2(y2 + z2) 0 z y ! = '#v 2á!'0c2(y0 2 + z0 2) 0 z0 y0 ! (5.18) In the second equality, we've rewritten the expression in terms of the coordinates of S0 which, because the boost is in the x-direction, are trivial: y = y0 and z = z0. From the perspective of an observer in frame S0, the charge density in the wire is '#0 = '#, where the factor of comes from Lorentz contraction.

This can be seen in the expression above for the electric field. Since the charge density is now moving, the observer in frame S0 sees a current I0 = '#v. Then we can rewrite (5.18) as B0 = µ0I0 2á! py0 2 + z0 2 '^ 0 (5.19) But this is something that we've seen before. It's the magnetic field due to a current in a wire (3.5). We computed this in Section 3.1.1 using Amp´ere's law. But here we've re-derived the same result without ever mentioning Amp´ere's law! Instead, our starting point (5.17) needed Gauss' law and we then used only the Lorentz transformation of electric and magnetic fields. We can only conclude that, under a Lorentz transformation, Gauss' law must be related to Amp´ere's law. Indeed, we'll shortly see explicitly that this is the case. For now, it's worth repeating the lesson that we learned in Section 5.2.1: the magnetic field can be viewed as a relativistic eµ!ect 5.3.4 Another Example: A Boosted Point Charge Consider a point charge Q, stationary in an inertial frame S. We know that it's electric field is given by E = Q 4á!'0r2 ^r = Q 4á!'0[x2 + y2 + z2] 3/2 x y z !while its magnetic field vanishes. Now let's look at this same particle from the frame S0, moving with velocity v = (v, 0, 0) with respect to S. The Lorentz boost which relates the two is given by (5.4) and so the new electric field are given by (5.15), E0 = Q 4á!'0[x2 + y2 + z2] 3/2 x y z ! But this is still expressed in terms of the original coordinates. We should now rewrite this in terms of the coordinates of S0, which are x0 = (x vt) and y0 = y and z0 = z. Inverting these, we have E0 = Q 4á!'0[2(x0 + vt0)2 + y0 2 + z0 2] 3/2 x0 + vt0

1. y0 z0 ! (5.20) In the frame S0, the particle sits at x0 = (vt0, 0, 0), so we see that the electric field emanates from the position of the charge, as it should. For now, let's look at the electric field when t 0 = 0 so that the particle sits at the origin in the new frame. The electric field points outwards radially, along the direction r0 = x0 y0 z0 ! However, the electric field is not isotropic. This arises from the denominator of (5.20) which is not proportional to r0 3 because there's an extra factor of 2 in front of the x0 component. Instead, at t 0 = 0, the denominator involves the combination 2 x0 2 + y0 2 + z0 2 = (2 1)x0 2 + r0 2 = v22 c2 x0 2 + r0 2 = –'v22 c2 cos2 –' + 1Æ% r0 2 = 2 –' 1 v2 c2 sin2 –'Æ% v Figure 46: The isotropic field lines of a static charge Figure 47: The squeezed field lines of a moving charge where the –' is the angle between r0 and the x0 -axis and, in the last line, we've just used some simple trig and the definition of 2 = 1/(1 v2/c2). This means that we can write the electric field in frame S0 as E0 = 1 2(1 v2 sin2 –'/c2)3/2 Q 4á!'0r0 2 ^r0 The pre-factor is responsible for the fact that the electric field is not isotropic. We see that it reduces the electric field along the x0 -axis (i.e when –' = 0) and increases the field along the perpendicular y0 and z0 axes (i.e. when –' = á!/2). This can be thought of as a consequence of Lorentz contraction, squeezing the electric field lines in the direction of travel. The moving particle also gives rise to a magnetic field. This is easily computed using the Lorentz transformations (5.16). It is B = µ0Qv 4á![2(x0 + vt0)2 + y0 2 + z0 2] 3/2 0 z0 y0 ! 5.3.5 Lorentz Scalars We can now ask a familiar question: is there any combination of the electric and magnetic fields that all observers agree upon? Now we have the power of index notation at our disposal, this is easy to answer. We just need to write down an object that doesn't have any floating µ or +# indices. Unfortunately, we don't get to use the obvious choice of '#µ+#F µ+# because this vanishes on account of the anti-symmetry of F µ+#. The simplest thing we can write down is 1 2 Fµ+#F µ+# = E2 c2 + B2 Note the relative minus sign between E and B, mirroring a similar minus sign in the spacetime interval. However, this isn't the only Lorentz scalar that we can construct from E and B.

There is another, somewhat more subtle, object. To build this, we need to appreciate that Minkowski spacetime comes equipped with another natural tensor object, beyond the familiar metric '#µ+#. This is the fully anti-symmetric object known as the alternating tensor, ' µ+#â! = (+1 if µ+#â! is an even permutation of 0123 1 if µ+#â! is an odd permutation of 0123 while 'µ+#â! = 0 if there are any repeated indices. To see why this is a natural object in Minkowski space, let's look at how it changes under Lorentz transformations. The usual tensor transformation is ' 0 µ+#â! = ä!µ ?ä!+#ä!â!µ!ä!' ?µ! It's simple to check that '0 µ+#â! is also full anti-symmetric; it inherits this property from '?µ! on the right-hand side. But this means that '0 µ+#â! must be proportional to 'µ+#â!. We only need to determine the constant of proportionality. To do this, we can look at ' 0 0123 = ä!0 ?ä!1 ä!2 µ!ä!3 ' ?µ! = det(ä!) Now any Lorentz transformations have det(ä!) = ±1. Those with det(ä!) = 1 make up the "proper Lorentz group" SO(1, 3). (This was covered in the Dynamics and Relativity notes). These proper Lorentz transformations do not include reflections or time reversal. We learn that the alternating tensor 'µ+#â! is invariant under proper Lorentz transformations. What it's really telling us is that Minkowski space comes with an oriented orthonormal basis. By lowering indices with the Minkowski metric, we can also construct the tensor 'µ+#â! which has '0123 = 1.

The alternating tensor allows us to construct a second tensor field, sometimes called the dual electromagnetic tenso·· (although "dual" is perhaps the most overused word in physics), F˜µ+# = 1 2 ' µ+#â!Fâ! = 0 BBBB@ 0 Bx By Bz Bx 0 Ez/c Ey/c By Ez/c 0 Ex/c Bz Ey/c Ex/c 0 1 CCCCA (5.21) F˜µ+# is sometimes also written as ?F µ+#. We see that this is looks just like F µ+# but with the electric and magnetic fields swapped around. Actually, looking closely you'll see that there's a minus sign diµ!erence as well: F˜µ+# arises from F µ+# by the substitutio The statement that F˜µ+# is a tensor means that it too has nice properties under Lorentz transformations, F˜0 µ+# = ä!µ â!ä!+# F˜â! and we can use this to build new Lorentz invariant quantities. Taking the obvious square of F˜ doesn't give us anything new, since F˜µ+#F˜µ+# = F µ+#Fµ+# But by contracting F˜ with the original F we do find a new Lorentz invariant 1 4 F˜µ+#Fµ+# = 1 c E · B This tells us that the inner-product of E and B is the same viewed in all frames. 5.4 Maxwell Equations We now have the machinery to write the Maxwell equations in a way which is manifestly compatible with special relativity.

They take a particularly simple form: @µF µ+# = µ0J+# and @µF˜µ+# = 0 (5.22) Pretty aren't they! The Maxwell equations are not invariant under Lorentz transformations. This is because there is the dangling +# index on both sides. However, because the equations are built out of objects which transform nicely – F µ+#, F˜µ+#, Jµ and @µ – the equations themselves also transform nicely. For example, we will see shortly that Gauss' law transforms into Amp´ere's law under a Lorentz boost, something we anticipated in Section 5.3.3. We say that the equations are covariant under Lorentz transformations. This means that an observer in a diµ!erent frame will mix everything up: space and time, charges and currents, and electric and magnetic fields. Although observers disagree on what these things are, they all agree on how they fit together. This is what it means for an equation to be covariant: the ingredients change, but the relationship between them stays the same. All observers agree that, in their frame, the electric and magnetic fields are governed by the same Maxwell equations. Given the objects F µ+#, F˜µ+#, Jµ and @µ, the Maxwell equations are not the only thing you could write down compatible with Lorentz invariance. But they are by far the simplest. Any other equation would be non-linear in F or F˜ or contain more derivative terms or some such thing. Of course, simplicity is no guarantee that equations are correct. For this we need experiment. But surprisingly often in physics we find that the simplest equations are also the right ones. – 11 Unpacking the Maxwell Equations Let's now check that the Maxwell equations (5.22) in relativistic form do indeed coincide with the vector calculus equations that we've been studying in this course. We just need to expand the diµ!erent parts of the equation.

The components of the first Maxwell equation give @iFi0 = µ0J0) r · E = â!'0 @µF µi = µ0Ji) 1 c2 @E @t + r å! B = µ0J In the first equation, which arises from +# = 0, we sum only over spatial indices i = 1, 2, 3 because F00 = 0. Meanwhile the components of the second Maxwell equation give @iF˜i0 = 0) r · B = 0 @µF˜µi = 0) @B @t + r å! E = 0 These, of course, are the familiar equations that we've all grown to love over this course. Here a few further, simple comments about the advantages of writing the Maxwell equations in relativistic form. First, the Maxwell equations imply that current is conserved. This follows because F µ+# is anti-symmetric, so @µ@+#F µ+# = 0 automatically, simply because @µ@+# is symmetric. The first of the Maxwell equations (5.22) then requires that the continuity equation holds @µJµ = 0 This is the same calculation that we did in vector notation in Section 4.2.1. Note that it's marginally easier in the relativistic framework. The second Maxwell equation can be written in a number of diµ!erent ways. It is equivalent to: @µF˜µ+# = 0, ' µ+#â!@+#Fâ! = 0, @â!Fµ+# + @+#Fâ!µ + @µF+#â! = 0 where the last of these equalities follows because the equation is constructed so that it is fully anti-symmetric with respect to exchanging any of the indices â!, µ and +#. (Just expand out for a few examples to see this). – 11

The gauge potential Aµ was originally introduced to solve the two Maxwell equations which are contained in @µF˜µ+# = 0. Again, this is marginally easier to see in relativistic notation. If we write Fµ+# = @µA+# @+#Aµ then @µF˜µ+# = 1 2 ' µ+#â!@µFâ! = 1 2 ' µ+#â!@µ(@â!A @Aâ!)=0 where the final equality holds because of the symmetry of the two derivatives, combined with the anti-symmetry of the '-tensor. This means that we could equally well write the Maxwell equations as @µF µ+# = µ0J+# where Fµ+# = @µA+# @+#Aµ In more advanced formulations of electromagnetism (for example, in the Lagrangian formulation), this is the form in which the Maxwell equations arise. 5.5 The Lorentz Force Law There's one last aspect of electromagnetism that we need to show is compatible with relativity: the Lorentz force law. In the Newtonian world, the equation of motion for a particle moving with velocity u and momentum p = mu is dp dt = q(E + u å! B) (5.23) We want to write this equation in 4-vector notation in a way that makes it clear how all the objects change under Lorentz transformations. By now it should be intuitively clear how this is going to work.

A moving particle experiences the magnetic force. But if we boost to its rest frame, there is no magnetic force. Instead, the magnetic field transforms into an electric field and we find the same force, now interpreted as an electric force. The relativistic version of (5.23) involves the 4-momentum P µ, defined in (5.5), the proper time '#, reviewed in Section 5.1.2, and our new friend the electromagnetic tensor F µ+#. The electromagnetic force acting on a point particle of charge q can then be written as dP µ d'# = q F µ+#U+#

(5.24) where the 4-velocity is Uμ = dXμ d′# = c u ! (5.25) and the 4-momentum is P = mU. Again, we see that the relativistic form of the equation (5.24) is somewhat prettier than the original equation (5.23). Unpacking the Lorentz Force Law Let's check to see that the relativistic equation (5.24) is giving us the right physics.

It is, of course, four equations: one for each μ = 0, 1, 2, 3. It's simple to multiply out the right-hand side, remembering that Uμ comes with an extra minus sign in the spatial components relative to (5.25). We find that the μ = 1, 2, 3 components of (5.24) arrange themselves into a familiar vector equation, dp d′# = q(E + u å! B)) dp dt = q(E + u å! B) (5.26) where we've used the relationship dt/d′# =. We find that we recover the Lorentz force law. Actually, there's a slight diμ!erence from the usual Newtonian force law (5.23), although the diμ!erence is buried in our notation. In the Newtonian setting, the momentum is p = mu. However, in the relativistic setting above, the momentum is p = mu. Needless to say, the relativistic version is correct, although the diμ!erence only shows up at high speeds. The relativistic formulation of the Lorentz force (5.24) also contains an extra equation coming from μ = 0. This reads dP0 d′# = q c E · u (5.27) Recall that the temporal component of the four-momentum is the energy P0 = E/c. Here the energy is E = mc2 which includes both the rest-mass of the particle and its kinetic energy. The extra equation in (5.24) is simply telling us that the kinetic energy increases when work is done by an electric field d(Energy) dt = qE · u where I've written energy as a word rather than as E to avoid confusing it with the electric field E. 5.5.1 Motion in Constant Fields We already know how electric and magnetic fields act on particles in a Newtonian world. Electric fields accelerate particles in straight lines; magnetic fields make particles go in circles. Here we're going to redo this analysis in the relativistic framework. The Lorentz force law remains the same. The only diμ!erence is that momentum is now p = mu. We'll see how this changes things. – Constant Electric Field Consider a vanishing magnetic field and constant electric field E = (E, 0, 0). (Note that E here denotes electric field, not energy!). The equation of motion (5.26) for a charged particle with velocity u = (u, 0, 0) is m d(u) dt = qE) mu = qEt where we've implicitly assumed that the particle starts from rest at t = 0. Rearranging, we get u = dx dt = qEt pm2 + q2E2t2/c2 Reassuringly, the speed never exceeds the speed of light. Instead, u !c as t ! 1 as one would expect. It's simple to integrate this once more. If the particle starts from the origin, we have x = mc2 qE r 1 + q2E2t2 m2c2 1 ! For early times, when the speeds are not too high, this reduces to mx á! 1 2 qEt2 +... which is the usual non-relativistic result for particles undergoing constant acceleration in a straight line. Constant Magnetic Field Now let's turn the electric field oμ! and look at the case of constant magnetic field B = (0, 0, B). In the non-relativistic world, we know that particles turn circles with frequency ! = qB/ m. Let's see how relativity changes things. We start by looking at the zeroth component of the force equation (5.27) which, in the absence of an electric field, reads dP0 d'# = 0 This tells us that magnetic fields do no work. We knew this from our course on Newtonian physics, but it remains true in the relativistic context. So we know that energy, E = mc2, is constant. But this tells us that the speed (i.e. the magnitude of – 1 the velocity) remains constant. In other words, the velocity, and hence the position, once again turn circles. The equation of motion is now m d(u) dt = qu å! B Since is constant, the equation takes the same form as in the non-relativistic case and the solutions are circles (or helices if the particle also moves in the z-direction). The only diµ!erence is that the frequency with which the particle moves in a circle now depends on how fast the particle is moving, ! = qB m If you wanted, you could interpret this as due to the relativistic increase in the mass of a moving particle. Naturally, for small speeds á! 1 and we reproduce the more familiar cyclotron frequency á! qB/m. So far we have looked at situations in which E = 0 and in which B = 0. But we've seen that E·B = 0 and E2B2 are both Lorentz invariant quantities. This means that the solutions we've described above can be boosted to apply to any situation where E · B = 0 and E2 B2 is either > 0 or < 0. In the general situation, both electric and magnetic fields are turned on so E · B 6= 0 and we have three possibilities to consider depending on whether E2 B2 is > 0 or < 0 or = 0.

THE PRINCIPLE FEYNMAN AND EINSTEIN COULD NOT FIND WAS DISCOVERED

Why Faraday's law and Lorentz force create the same electromotive force?

The Faraday's induction formula (flux rule) of electromagnetism says that the electromotive force (emf) created in a conducting circuit is equal to the rate at which the magnetic flux through the conducting circuit changes as it is written on a high school text in physics. This emf can be calculated in two ways: either by using the Lorentz force formula and calculating the force acting on electrons in the moving conductor of the circuit; or via one of Maxwell's equations (Faraday's law) and calculating the change of the magnetic flux penetrating through the circuit. The Lorentz force formula and Maxwell's equations are two distinct physical laws, yet the two methods yield the same results.

The Faraday's induction formula (flux rule) of electromagnetism says that the electromotive force (emf) created in a conducting circuit is equal to the rate at which the magnetic flux through the conducting circuit changes as it is written on a high school text in physics. This emf can be calculated in two ways: either by using the Lorentz force formula and calculating the force acting on electrons in the moving conductor of the circuit; or via one of Maxwell's equations (Faraday's law) and calculating the change of the magnetic flux penetrating through the circuit. The Lorentz force formula and Maxwell's equations are two distinct physical laws, yet the two methods yield the same results.

Why the two results coincide was not known. In other words, the flux rule consists of two physically different laws in classical theories. Interestingly, this problem was also a motivation behind the development of the theory of relativity by Albert Einstein. In 1905, Einstein wrote in the opening paragraph of his first paper on relativity theory, "It is known that Maxwell's electrodynamics as usually understood at the present time when applied to moving bodies, leads to asymmetries which do not appear to be inherent in the phenomena." But Einstein's argument moved away from this problem and formulated special theory of relativity, thus the problem was not solved.

Richard Feynman once described this situation in his famous lecture (The Feynman Lectures on Physics, Vol. II, 1964), "we know of no other place in physics where such a simple and accurate general principle requires for its real understanding an analysis in terms of two different phenomena. Usually such a beautiful generalization is found to stem from a single deep underlying principle. eÿeÿeÿeÿeÿeÿWe have to understand the "rule" as the combined effects of two quite separate phenomena."

Dr. Hiroyasu Koizumi's study, recently published in the Journal of Superconductivity and Novel Magnetism, has revealed what is the "single deep underlying principle" in the "flux rule" envisaged by Feynman. It is a duality of the U(1) phase factor added on the wave function; it describes the translational motion of electrons, and also gives a time-dependent gauge potential that induces an effective electric field on the electrons. The former view corresponds to the result obtained by the Lorentz force formula, and the latter to the result using the Maxwell's equation for the Faraday's law.

Behind this discovery are two big developments in physics in the 20th century. One is the birth of quantum mechanics, and the other is the establishment of the physical reality of the electromagnetic field as a U(1) gauge field. In the above study, electrons in the conductor are described by the wave functions of quantum mechanics and the magnetic field is expressed as the U(1) gauge field. The gauge field has an arbitrariness called the gauge degree-of-freedom. This arbitrariness can be cast in the U(1) phase factor on the wave function, and can be fixed by the requirement of the energy to be minimum. Then, the duality that the U(1) phase factor can be added to the wave function as the translational motion of electrons allows the "time-dependent gauge potential" to emerge. The same gauge fixing has been employed in Dr. Koizumi's study on superconductivity, where the gauge fixing is achieved by the energy minimum requirement under the constraint that the wave function be a single-valued function of the electron coordinates.

Dr. Koizumi's work also provides new perspectives on superconductivity and possibly also string theory. Since the most promising qubits for quantum computers are now those using superconductors, the present finding is expected to contribute to the development of quantum computers that may supersede classical computers.

EINSTEIN'S PHILOSOPHY OF SCIENCE

FIRST PUBLISHED WED FEB 11, 2004

Albert Einstein (1879–1955) is well known as the most prominent physicist of the twentieth century. Less well known, though of comparable importance, are his contributions to twentieth-century philosophy of science. Einstein's own philosophy of science is an original synthesis of elements drawn from sources as diverse as neo-Kantianism, conventionalism, and logical empiricism, its distinctive feature being its novel blending of realism with a holist, underdeterminationist form of conventionalism. Of special note is the manner in which Einstein's philosophical thinking was driven by and contributed to the solution of problems first encountered in his work in physics. Equally significant are Einstein's relations with and influence on other prominent twentieth-century philosophers of science, including Philipp Frank, Moritz Schlick, Ernst Cassirer, Henri Bergson, and Hans Reichenbach.

- Introduction: Was Einstein an Epistemological "Opportunist"?
- The Underdetermination of Theory Choice by Evidence: The Nature and Role of Conventions in Science
- Simplicity and Theory Choice
- Univocalness in the Theoretical Representation of Nature
- Realism and Separability
- The Principle Theories—Constructive Theories Distinction
- Conclusion: Albert Einstein: Philosopher-Physicist
- Bibliography
- Einstein's Work
- Related Literature
- Academic Tools
- Other Internet Resources
- Related Entries.

INTRODUCTION: WAS EINSTEIN AN EPISTEMOLOGICAL "OPPORTUNIST"?

Late in 1944, Albert Einstein received a letter from Robert Thornton, a young African-American philosopher of science who had just finished his Ph.D. under Herbert Feigl at Minnesota and was beginning a new job teaching physics at the University of Puerto Rico, Mayaguez. He had written to solicit from Einstein a few supportive words on behalf of his efforts to introduce "as much of the philosophy of science as possible" into the modern physics course that he was to teach the following spring (Thornton to Einstein, 28 November 1944, EA 61–573). Here is what Einstein offered in reply:

I fully agree with you about the significance and educational value of methodology as well as history and philosophy of science. So many people today—and even professional scientists—seem to me like somebody who has seen thousands of trees but has never seen a forest. A knowledge of the historic and philosophical background gives that kind of independence from prejudices of his generation from which most scientists are suffering. This independence created by philosophical insight is—in my opinion—the mark of distinction between a mere artisan or specialist and a real seeker after truth. (Einstein to Thornton, 7 December 1944, EA 61-574)

That Einstein meant what he said about the relevance of philosophy to physics is evidenced by the fact that he had been saying more or less the same thing for decades. Thus, in a 1916 memorial note for Ernst Mach, a physicist and philosopher to whom Einstein owed a special debt, he wrote:

How does it happen that a properly endowed natural scientist comes to concern himself with epistemology? Is there no more valuable work in his specialty? I hear many of my colleagues saying, and I sense it from many more, that they feel this way. I cannot share this sentiment. When I think about the ablest students whom I have encountered in my teaching, that is, those who distinguish themselves by their independence of judgment and not merely their quick-wittedness, I can affirm that they had a vigorous interest in epistemology. They happily began discussions about the goals and methods of science, and they showed unequivocally, through their tenacity in defending their views, that the subject seemed important to them. Indeed, one should not be surprised at this. (Einstein 1916, 101)

How, exactly, does the philosophical habit of mind provide the physicist with such "independence of judgment"? Einstein goes on to explain:

Concepts that have proven useful in ordering things easily achieve such an authority over us that we forget their earthly origins and accept them as unalterable givens. Thus they come to be stamped as "necessities of thought," "a priori givens," etc. The path of scientific advance is often made impassable for a long time through such errors. For that reason, it is by no means an idle game if we become practiced in analyzing the long commonplace concepts and exhibiting those circumstances upon which their justification and usefulness depend, how they have grown up, individually, out of the givens of experience. By this means, their all-too-great authority will be broken. They will be removed if they cannot be properly legitimated, corrected if their correlation with given things be far too superfluous, replaced by others if a new system can be established that we prefer for whatever reason. (Einstein 1916, 102)

One is not surprised at Einstein's then citing Mach's critical analysis of the Newtonian conception of absolute space as a paradigm of what Mach, himself, termed the "historical-critical" method of philosophical analysis (Einstein 1916, 101, citing Ch. 2, §§ 6–7 of Mach's Mechanik, most likely the third edition, Mach 1897).

The place of philosophy in physics was a theme to which Einstein returned time and again, it being clearly an issue of deep importance to him. Sometimes he adopts a modest pose, as in this oft-quoted remark from his

1933 Spencer Lecture: If you wish to learn from the theoretical physicist anything about the methods which he uses, I would give you the following piece of advice: Don't listen to his words, examine his achievements. For to the discoverer in that field, the constructions of his imagination appear so necessary and so natural that he is apt to treat them not as the creations of his thoughts but as given realities. (Einstein 1933, 5–6)

More typical, however, is the confident pose he struck three years later in "Physics and Reality":

It has often been said, and certainly not without justification, that the man of science is a poor philosopher. Why then should it not be the right thing for the physicist to let the philosopher do the philosophizing? Such might indeed be the right thing at a time when the physicist believes he has at his disposal a rigid system of fundamental concepts and fundamental laws which are so well established that waves of doubt can not reach them; but it can not be right at a time when the very foundations of physics itself have become problematic as they are now. At a time like the present, when experience forces us to seek a newer and more solid foundation, the physicist cannot simply surrender to the philosopher the critical contemplation of the theoretical foundations; for, he himself knows best, and feels more surely where the shoe pinches. In looking for a new foundation, he must try to make clear in his own mind just how far the concepts which he uses are justified, and are necessities. (Einstein 1936, 349)

What kind of philosophy might we expect from the philosopher-physicist? One thing that we should not expect from a physicist who takes the philosophical turn in order to help solve fundamental physical problems is a systematic philosophy:

The reciprocal relationship of epistemology and science is of noteworthy kind. They are dependent upon each other. Epistemology without contact with science becomes an empty scheme. Science without epistemology is-insofar as it is thinkable at all-primitive and muddled. However, no sooner has the epistemologist, who is seeking a clear system, fought his way through to such a system, than he is inclined to interpret the thought-content of science in the sense of his system and to reject whatever does not fit into his system. The scientist, however, cannot afford to carry his striving for epistemological systematic that far. He accepts gratefully the epistemological conceptual analysis; but the external conditions, which are set for him by the facts of experience, do not permit him to let himself be too much restricted in the construction of his conceptual world by the adherence to an epistemological system. He therefore must appear to the systematic epistemologist as a type of unscrupulous opportunist: he appears as realist insofar as he seeks to describe a world independent of the acts of perception; as idealist insofar as he looks upon the concepts and theories as free inventions of the human spirit (not logically derivable from what is empirically given); as positivist insofar as he considers his concepts and theories justified only to the extent to which they furnish a logical representation of relations among sensory experiences. He may even appear as Platonist or Pythagorean insofar as he considers the viewpoint of logical simplicity as an indispensable and effective tool of his research. (Einstein 1949, 683-684)

But what strikes the "systematic epistemologist" as mere opportunism might appear otherwise when viewed from the perspective of a physicist engaged, as Einstein himself put it, in "the critical contemplation of the theoretical foundations." The overarching goal of that critical contemplation was, for Einstein, the creation of a unified foundation for physics after the model of a field theory like general relativity. Einstein failed in his quest, but there was a consistency and constancy in the striving that informed as well the philosophy of science developing hand in hand with the scientific project.

Indeed, from early to late a few key ideas played the central, leading role in Einstein's philosophy of science, ideas about which Einstein evinced surprisingly little doubt even while achieving an ever deeper understanding of their implications. For the purposes of the following comparatively brief overview, we can confine our attention to just five topics:

- The underdetermination of theory choice by evidence.
- Simplicity and theory choice.
- Univocalness in the theoretical representation of nature.
- Realism and separability.
- The principle theories-constructive theories distinction.

In emphasizing the continuity and coherence in the development of Einstein's philosophy of science, I take issue with an account such as Gerald Holton's (1968), which claims to find a major philosophical break in the mid-1910s, in the form of a turn away from a sympathy for an anti-metaphysical positivism and toward a robust scientific realism. Holton sees this turn being driven by Einstein's alleged realization that general relativity, by contrast with special relativity, requires a realistic ontology. On my view, Einstein was never an ardent "Machian" positivist, and he was never a scientific realist, at least not in the sense acquired by the term "scientific realist" in later twentieth century philosophical discourse (see Howard 1993). Einstein expected scientific theories to have the proper empirical credentials, but he was no positivist; and he expected scientific theories to give an account of physical reality, but he was no scientific realist. Moreover, in both respects his views remained more or less the same from the beginning to the end of his career.

Why Einstein did not think himself a realist (he said so explicitly) is discussed below. Why he is not to be understood as a positivist deserves a word or two of further discussion here, if only because the belief that he was sympathetic to positivism, at least early in his life, is so widespread (for a fuller discussion, see Howard 1993).

That Einstein later repudiated positivism is beyond doubt. Many remarks from at least the early 1920s through the end of his life make this clear.

In 1946 he explained what he took to be Mach's basic error: He did not place in the correct light the essentially constructive and speculative nature of all thinking and more especially of scientific thinking; in consequence, he condemned theory precisely at those points where its constructive-speculative character comes to light unmistakably, such as in the kinetic theory of atoms. (Einstein 1946, 21)

Is Einstein here also criticizing his own youthful philosophical indiscretions? The very example that Einstein gives here makes any such interpretation highly implausible, because one of Einstein's main goals in his early work on Brownian motion (Einstein 1905b) was precisely to prove the reality of atoms, this in the face of the then famous skepticism of thinkers like Mach and Wilhelm Ostwald:

My principal aim in this was to find facts that would guarantee as much as possible the existence of atoms of definite size.... The agreement of these considerations with experience together with Planck's determination of the true molecular size from the law of radiation (for high temperatures) convinced the skeptics, who were quite numerous at that time (Ostwald, Mach), of the reality of atoms. (Einstein 1946, 45, 47)

Why, then, is the belief in Einstein's early sympathy for positivism so well entrenched?

The one piece of evidence standardly cited for a youthful flirtation with positivism is Einstein's critique of the notion of absolute distant simultaneity in his 1905 paper on special relativity (Einstein 1905c). Einstein speaks there of "observers," but in an epistemologically neutral way that can be replaced by talk of an inertial frame of reference. What really bothers Einstein about distant simultaneity is not that it is observationally inaccessible but that it involves a two-fold arbitrariness, one in the choice of an inertial frame of reference and one in the stipulation within a given frame of a convention regarding the ratio of the times required for a light signal to go from one stationary observer to another and back again. Likewise, Einstein faults classical Maxwellian electrodynamics for an asymmetry in the way it explains electromagnetic induction depending on whether it is the coil or the magnet that is assumed to be at rest. If the effect is the same-a current in the coil-why, asks Einstein, should there be two different explanations: an electrical field created in the vicinity of a moving magnet or an electromotive force induced in a conductor moving through a stationary magnetic field? To be sure, whether it is the coil or the magnet that is taken to be at rest makes no observable difference, but the problem, from Einstein's point of view, is the asymmetry in the two explanations. Even the young Einstein was no positivist.

First generation logical empiricists sought to legitimate their movement in part by claiming Einstein as a friend. They may be forgiven their putting a forced interpretation on arguments taken out of context. We can do better.

Einstein's philosophy of science is an original synthesis drawing upon many philosophical resources, from neo-Kantianism to Machian empiricism and Duhemian conventionalism. Other thinkers and movements, most notably the logical empiricists, drew upon the same resources. But Einstein put the pieces together in a manner importantly different from Moritz Schlick, Hans Reichenbach, and Rudolf Carnap, and he argued with them for decades about who was right (however much they obscured these differences in representing Einstein publicly as a friend of logical empiricism and scientific philosophy). Understanding how Einstein puts those pieces together therefore sheds light not only on the philosophical aspect of his own achievements in physics but also upon the larger history of the development of the philosophy of science in the twentieth century.

THE UNDERDETERMINATION OF THEORY CHOICE BY EVIDENCE: THE NATURE AND ROLE OF CONVENTIONS IN SCIENCE

Any philosophy of science must include an account of the relation between theory and evidence. Einstein learned about the historicity of scientific concepts from Mach. But his preferred way of modeling the logical relationship between theory and evidence was inspired mainly by his reading of Pierre Duhem's La Théorie physique: son objet et sa structure (Duhem 1906). Einstein probably first read Duhem, or at least learned the essentials of Duhem's philosophy of science around the fall of 1909, when, upon returning to Zurich from the patent office in Bern to take up his first academic appointment at the University of Zurich, he became the upstairs neighbor of his old friend and fellow Zurich physics student, Friedrich Adler. Just a few months before, Adler had published the German translation of La Théorie physique (Duhem 1908), and the philosophy of science became a frequent topic of conversation between the new neighbors, Adler and Einstein (see Howard 1990a).

Theoretical holism and the underdetermination of theory choice by empirical evidence are the central theses in Duhem's philosophy of science. His argument, in brief, is that at least in sciences like physics, where experiment is dense with sophisticated instrumentation whose employment itself requires theoretical interpretation, hypotheses are not tested in isolation but only as part of whole bodies of theory. It follows that when there is a conflict between theory and evidence, the fit can be restored in a multiplicity of different ways. No statement is immune to revision because of a presumed status as a definition or thanks to some other a priori warrant, and most any statement can be retained on pain of suitable adjustments elsewhere in the total body of theory. Hence, theory choice is underdetermined by evidence.

That Einstein's exposure to Duhem's philosophy of science soon left its mark is evident from lecture notes that Einstein prepared for a course on electricity and magnetism at the University of Zurich in the winter semester of 1910/11. Einstein asks how one can assign a definite electrical charge everywhere within a material body, if the interior of the body is not accessible to test particles. A "Machian" positivist would deem such direct empirical access necessary for meaningful talk of a charge distribution in the interior of a sold. Einstein argues otherwise: We have seen how experience led to the introd. of the concept of the quantity of electricity. it was defined by means of the forces that small electrified bodies exert on each other. But now we extend the application of the concept to cases in which this definition cannot be applied directly as soon as we conceive the el. forces as forces exerted on electricity rather than on material particles. We set up a conceptual system the individual parts of which do not correspond directly to empirical facts. Only a certain totality of theoretical material corresponds again to a certain totality of experimental facts.

We find that such an el. continuum is always applicable only for the representation of el. states of affairs in the interior of ponderable bodies. Here too we define the vector of el. field strength as the vector of the mech. force exerted on the unit of pos. electr. quantity inside a body. But the force so defined is no longer directly accessible to exp. It is one part of a theoretical construction that can be correct or false, i.e., consistent or not consistent with experience, only as a whole. (EA 3-007, ECP 3-11, 325)

One can hardly ask for a better summary of Duhem's point of view in application to a specific physical theory.

Explicit citations of Duhem by Einstein are rare (for details, see Howard 1990a). But explicit invocations of a holist picture of the structure and empirical interpretation of theories are not hard to find. An especially interesting example is found in a review that Einstein wrote in 1924 of Alfred Elsbach's Kant und Einstein (1924), one of the flood of books and articles then trying to reconcile the Kantian doctrine of the a priori Euclidean character of space with general relativity's postulate of variable spacetime curvature. Having asserted that relativity theory is incompatible with Kant's doctrine of the a priori, Einstein explains why, more generally, he is not sympathetic with Kant:

This does not, at first, preclude one's holding at least to the Kantian problematic, as, e.g., Cassirer has done. I am even of the opinion that this standpoint can be rigorously refuted by no development of natural science. For one will always be able to say that critical philosophers have until now erred in the establishment of the a priori elements, and one will always be able to establish a system of a priori elements that does not contradict a given physical system. Let me briefly indicate why I do not find this standpoint natural. A physical theory consists of the parts (elements) A, B, C, D, that together constitute a logical whole which correctly connects the pertinent experiments (sense experiences). Then it tends to be the case that the aggregate of fewer than all four elements, e.g., A, B, D, without C, no longer says anything about these experiences, and just as well A, B, C without D. One is then free to regard the aggregate of three of these elements, e.g., A, B, C as a priori, and only D as empirically conditioned. But what remains unsatisfactory in this is always the arbitrariness in the choice of those elements that one designates as a priori, entirely apart from the fact that the theory could one day be replaced by another that replaces certain of these elements (or all four) by others. (Einstein 1924, 1688-1689)

Einstein's point seems to be that while one can always choose to designate selected elements as a priori and, hence, non-empirical, no principle determines which elements can be so designated, and our ability thus to designate them derives from the fact that it is only the totality of the elements that possesses empirical content.

Much the same point could be made, and was made by Duhem himself (see Duhem 1906, part 2, ch. 6, sects. 8 and 9), against those who would insulate certain statements against empirical refutation by claiming for them the status of conventional definitions. Edouard Le Roy (1901) had argued thus about the law of free fall. It could not be refuted by experiment because it functioned as a definition of "free fall." And Henri Poincaré (1901) said much the same about the principles of mechanics more generally. As Einstein answered the neo-Kantians, so Duhem answered this species of conventionalist: Yes, experiment cannot refute, say, the law of free fall by itself, but only because it is part of a larger theoretical whole that has empirical content only as a whole, and various other elements of that whole could as well be said to be, alone, immune to refutation.

That Einstein should deploy against the neo-Kantians in the early 1920s the argument that Duhem used against the conventionalism of Poincaré and Le Roy is interesting from the point of view of Einstein's relationships with those who were leading the development of logical empiricism and scientific philosophy in the 1920s, especially Schlick and Reichenbach. Einstein shared with Schlick and Reichenbach the goal of crafting a new form of empiricism that would be adequate to the task of defending general relativity against neo-Kantian critiques (see Schlick 1917 and 1921, and Reichenbach 1920, 1924, and 1928; for more detail, see Howard 1994a). But while they all agreed that what Kant regarded as the a priori element in scientific cognition was better understood as a conventional moment in science, they were growing to disagree dramatically over the nature and place of conventions in science. With the help of new logical tools and a more sophisticated verificationist semantics, Schlick and Reichenbach were refining Poincaré's idea of conventional definitional elements in science into the classic logical empiricist view that the moment of convention was restricted to conventional coordinating definitions that endow individual primitive terms and, by extension, the individual synthetic propositions constructed out of them with empirical content. This view worked well as an answer to the neo-Kantian, for it implied that once one fixed one's coordinating definition-as with a conventional choice of a standard measuring rod coordinated with the geometer's concept of a "rigid body"-the question of the curvature of space had an empirically determinate answer. But unless the division is wholly arbitrary, parsing theories thus into coordinating definitions and empirical statements assumes a principled difference in kind between the two categories of statements along the lines of an analytic-synthetic distinction. As had been the case with Duhem before him, the assumption of such a principled difference in kind did not comport well with the holism about theories that Einstein had learned from Duhem.

It was this argument over the nature and place of conventions in science that underlay Einstein's gradual philosophical estrangement from Schlick and Reichenbach in the 1920s. Serious in its own right, the argument over conventions was entangled with two other issues as well, namely, realism and Einstein's famous view of theories as the "free creations of the human spirit" (see, for example, Einstein 1921). In both instances what troubled Einstein was that a verificationist semantics made the link between theory and experience too strong, leaving too small a role for theory, itself, and the creative theorizing that produces it.

If theory choice is empirically determinate, especially if theoretical concepts are explicitly constructed from empirical primitives, as in Carnap's program in the Aufbau (Carnap 1928), then it is hard to see how theory gives us a story about anything other than experience. As noted, Einstein was not what we would today call a scientific realist, but he still believed that there was content in theory beyond mere empirical content. He believed that theoretical science gave us a window on nature itself, even if, in principle, there will be no one uniquely correct story at the level of deep ontology (see below, section 5). And if the only choice in theory choice is one among conventional coordinating definitions, then that is no choice at all, a point stressed by Reichenbach, especially, as an important positive implication of his position. Reichenbach argued that if empirical content is the only content, then empirically equivalent theories have the same content, the difference resulting from their different choices of coordinating definitions being like in kind to the difference between "es regnet" and "il pleut," or the difference between expressing the result of a measurement in English or metric units, just two different ways of saying the same thing. But then, Einstein would ask, where is there any role for the creative intelligence of the theoretical physicist if there is no room for genuine choice in science, if experience somehow dictates theory construction?

The argument over the nature and role of conventions in science continued to the very end of Einstein's life, reaching its highest level of sophistication in the exchange between Reichenbach and Einstein the Library of Living Philosopher's volume, Albert Einstein: Philosopher-Physicist(Schilpp 1949). The question is, again, whether the choice of a geometry is empirical, conventional, or a priori. In his contribution, Reichenbach reasserted his old view that once an appropriate coordinating definition is established, equating some "practically rigid rod" with the geometer's "rigid body," then the geometry of physical space is wholly determined by empirical evidence:

The choice of a geometry is arbitrary only so long as no definition of congruence is specified. Once this definition is set up, it becomes an empirical question which geometry holds for physical space.... The conventionalist overlooks the fact that only the incomplete statement of a geometry, in which a reference to the definition of congruence is omitted, is arbitrary. (Reichenbach 1949, 297)

Einstein's clever reply includes a dialogue between two characters, "Reichenbach" and "Poincaré," in which "Reichenbach" concedes to "Poincaré" that there are no perfectly rigid bodies in nature and that physics must be used to correct for such things as thermal deformations, from which it follows that what we actually test is geometry plus physics, not geometry alone. Here an "anonymous non-positivist" takes "Poincaré's" place, out of respect, says Einstein, "for Poincaré's superiority as thinker and author" (Einstein 1949, 677), but also, perhaps, because he realized that the point of view that follows was more Duhem than Poincaré. The "non-positivist" then argues that one's granting that geometry and physics are tested together contravenes the positivist identification of meaning with verifiability:

Non-Positivist: If, under the stated circumstances, you hold distance to be a legitimate concept, how then is it with your basic principle (meaning = verifiability)? Must you not come to the point where you deny the meaning of geometrical statements and concede meaning only to the completely developed theory of relativity (which still does not exist at all as a finished product)? Must you not grant that no "meaning" whatsoever, in your sense, belongs to the individual concepts and statements of a physical theory, such meaning belonging instead to the whole system insofar as it makes "intelligible" what is given in experience? Why do the individual concepts that occur in a theory require any separate justification after all, if they are indispensable only within the framework of the logical structure of the theory, and if it is the theory as a whole that stands the test? (Einstein 1949, 678).

Two years before the Quine's publication of "Two Dogmas of Empiricism" (1951), Einstein here makes explicit the semantic implications of a thoroughgoing holism.

If theory choice is empirically underdetermined, then an obvious question is why we are so little aware of the underdetermination in the day-to-day conduct of science. In a 1918 address celebrating Max Planck's sixtieth birthday, Einstein approached this question via a distinction between practice and principle:

The supreme task of the physicist is ... the search for those most general, elementary laws from which the world picture is to be obtained through pure deduction. No logical path leads to these elementary laws; it is instead just the intuition that rests on an empathic understanding of experience. In this state of methodological uncertainty one can think that arbitrarily many, in themselves equally justified systems of theoretical principles were possible; and this opinion is, in principle, certainly correct. But the development of physics has shown that of all the conceivable theoretical constructions a single one has, at any given time, proved itself unconditionally superior to all others. No one who has really gone deeply into the subject will deny that, in practice, the world of perceptions determines the theoretical system unambiguously, even though no logical path leads from the perceptions to the basic principles of the theory. (Einstein 1918, 31; my translation)

But why is theory choice, in practice, seemingly empirically determined? Einstein hinted at an answer the year before in a letter to Schlick, where he commended Schlick's argument that the deep elements of a theoretical ontology have as much claim to the status of the real as do Mach's elements of sensation (Schlick 1917), but suggested that we are nonetheless speaking of two different kinds of reality. How do they differ?

It appears to me that the word "real" is taken in different senses, according to whether impressions or events, that is to say, states of affairs in the physical sense, are spoken of.

If two different peoples pursue physics independently of one another, they will create systems that certainly agree as regards the impressions ("elements" in Mach's sense). The mental constructions that the two devise for connecting these "elements" can be vastly different. And the two constructions need not agree as regards the "events"; for these surely belong to the conceptual constructions. Certainly on the "elements," but not the "events," are real in the sense of being "given unavoidably in experience."

But if we designate as "real" that which we arrange in the space-time-schema, as you have done in the theory of knowledge, then without doubt the "events," above all, are real.... I would like to recommend a clean conceptual distinction here. (Einstein to Schlick, 21 May 1917, EA 21-618, ECP 8-343)

Why, in practice, are physicists unaware of underdetermination? It is because ours is not the situation of "two different peoples pursu[ing] physics independently of one another." Though Einstein does not say it explicitly, the implication seems to be that apparent determination in theory choice is mainly a consequence of our all being similarly socialized as we become members of a common scientific community. Part of what it means to be a member of a such a community is that we have been taught to make our theoretical choices in accord with criteria or values that we hold in common.

SIMPLICITY AND THEORY CHOICE

For Einstein, as for many others, simplicity is the criterion that mainly steers theory choice in domains where experiment and observation no longer provide an unambiguous guide. This, too, is a theme sounded early and late in Einstein's philosophical reflections (for more detail, see Howard 1998, Norton 2000, and van Dongen 2002). For example, the just-quoted remark from 1918 about the apparent determination of theory choice in practice, contrasted with in-principle underdetermination continues:

Furthermore this conceptual system that is univocally coordinated with the world of experience is reducible to a few basic laws from which the whole system can be developed logically. With every new important advance the researcher here sees his expectations surpassed, in that those basic laws are more and more simplified under the press of experience. With astonishment he sees apparent chaos resolved into a sublime order that is to be attributed not to the rule of the individual mind, but to the constitution of the world of experience; this is what Leibniz so happily characterized as "pre-established harmony." Physicists strenuously reproach many epistemologists for their insufficient appreciation of this circumstance. Herein, it seems to me, lie the roots of the controversy carried on some years ago between Mach and Planck. (Einstein 1918, p. 31)

There is more than a little autobiography here, for as Einstein stressed repeatedly in later years, he understood the success of his own quest for a general theory of relativity as a result of his seeking the simplest set of field equations satisfying a given set of constraints.

Always a leitmotif, Einstein's celebration of simplicity as a guide to theory choice comes clearly to the fore in the early 1930s. Why then? The reason might well be that his faith in simplicity had been vindicated when, seemingly with a sigh of relief, he found that he could drop from the gravitational field equations the cosmological constant that he had introduced in 1917 for the purpose of blocking non-static solutions, for the introduction of the cosmological constant in the first place had represented to him "a considerable renunciation of the logical simplicity of the theory" (Einstein 1949, 684-685). That his faith in simplicity was reaffirmed is clear. Witness what he wrote in his 1933 Herbert Spencer lecture:

If, then, it is true that the axiomatic foundation of theoretical physics cannot be extracted from experience but must be freely invented, may we ever hope to find the right way? Furthermore, does this right way exist anywhere other than in our illusions? May we hope to be guided safely by experience at all, if there exist theories (such as classical mechanics) which to a large extent do justice to experience, without comprehending the matter in a deep way?

To these questions, I answer with complete confidence, that, in my opinion, the right way exists, and that we are capable of finding it. Our experience hitherto justifies us in trusting that nature is the realization of the simplest that is mathematically conceivable. I am convinced that purely mathematical construction enables us to find those concepts and those lawlike connections between them that provide the key to the understanding of natural phenomena. Useful mathematical concepts may well be suggested by experience, but in no way can they be derived from it. Experience naturally remains the sole criterion of the usefulness of a mathematical construction for physics. But the actual creative principle lies in mathematics. Thus, in a certain sense, I take it to be true that pure thought can grasp the real, as the ancients had dreamed. (Einstein 1933, p. 183; my translation)

Another consideration reinforcing Einstein's conviction that the theoretical physicist must trust simplicity is that physics is moving steadily into domains ever further removed from direct contact with observation and experiment. Before the 1960s, general relativity, itself, rested on a famously thin empirical footing, and empirical evidence provided even less of a guide in Einstein's search for a unified field theory. One year after the Herbert Spencer lecture, at a time when he was immersed in work on unified field theory, Einstein wrote:

The theory of relativity is a beautiful example of the basic character of the modern development of theory. That is to say, the hypotheses from which one starts become ever more abstract and more remote from experience. But in return one comes closer to the preeminent goal of science, that of encompassing a maximum of empirical contents through logical deduction with a minimum of hypotheses or axioms. The intellectual path from the axioms to the empirical contents or to the testable consequences becomes, thereby, ever longer and more subtle. The theoretician is forced, ever more, to allow himself to be directed by purely mathematical, formal points of view in the search for theories, because the physical experience of the experimenter is not capable of leading us up to the regions of the highest abstraction. Tentative deduction takes the place of the predominantly inductive methods appropriate to the youthful state of science. Such a theoretical structure must be quite thoroughly elaborated in order for it to lead to consequences that can be compared with experience. It is certainly the case that here, as well, the empirical fact is the all-powerful judge. But its judgment can be handed down only on the basis of great and difficult intellectual effort that first bridges the wide space between the axioms and the testable consequences. The theorist must accomplish this Herculean task with the clear understanding that this effort may only be destined to prepare the way for a death sentence for his theory. One should not reproach the theorist who undertakes such a task by calling him a fantast; instead, one must allow him his fantasizing, since for him there is no other way to his goal whatsoever. Indeed, it is no planless fantasizing, but rather a search for the logically simplest possibilities and their consequences. (Einstein 1954, 238-239; my translation)

What warrant is there for thus trusting in simplicity? At best one can do a kind of meta-induction. That "the totality of all sensory experience can be 'comprehended' on the basis of a conceptual system built on premises of great simplicity" will be derided by skeptics as a "miracle creed," but, Einstein adds, "it is a miracle creed which has been borne out to an amazing extent by the development of science" (Einstein 1950, p. 342).

But for all that Einstein's faith in simplicity was strong, he despaired of giving a precise, formal characterization of how we assess the simplicity of a theory. In 1946 he wrote about the perspective of simplicity (here termed the "inner perfection" of a theory):

This point of view, whose exact formulation meets with great difficulties, has played an important role in the selection and evaluation of theories from time immemorial. The problem here is not simply one of a kind of enumeration of the logically independent premises (if anything like this were at all possible without ambiguity), but one of a kind of reciprocal weighing of incommensurable qualities.... I shall not attempt to excuse the lack of precision of [these] assertions ... on the grounds of insufficient space at my disposal; I must confess herewith that I cannot at this point, and perhaps not at all, replace these hints by more precise definitions. I believe, however, that a sharper formulation would be possible. In any case it turns out that among the "oracles" there usually is agreement in judging the "inner perfection" of the theories and even more so concerning the degree of "external confirmation." (Einstein 1946, pp. 21, 23).

As in 1918, so in 1946 and beyond, Einstein continues to be impressed that the "oracles," presumably the leaders of the relevant scientific community, tend to agree in their judgments of simplicity. That is why, in practice, simplicity seems to determine theory choice univocally.

Experience justifies our trusting that nature is the realization of the simplest that is mathematically conceivable. Can we say anything more about why this might be so? A hint is provided by Einstein's enthusiastically positive response to Schlick's first essay on the philosophical significance of relativity (Schlick 1915). At this early stage in his philosophical career, Schlick regarded himself as a realist and defended a version of the underdetermination thesis grounded in his view of truth as the unambiguous many-to-one coordination of propositions to facts (Schlick 1910). Theories being sets of propositions, several theories could likewise be unambiguously coordinated with a given set of facts and thus count as equally true representations of those facts. When he took up the question of simplicity, Schlick derided those who would justify simplicity as a criterion of theory choice by arguing that we should choose simple theories because nature itself is simple. As Schlick rightly pointed out, this is a circular argument, for our only cognitive access to nature is via our theories (note that Einstein argues not that nature, itself, is simple, but that nature is a realization of simple theoretical constructions, a crucial difference). Schlick similarly derides a defense grounded in considerations of mental economy, which he terms "intellectual indolence." Why then choose simple theories? Schlick's answer is that "the greater simplicity of a theory depends on its containing fewer arbitrary elements." Why is it better to choose theories with fewer "superfluous," arbitrary elements? Because only the non-arbitrary elements are likely to correspond to reality, so in choosing the simpler theory "we are then sure of diverging from reality at least no further than is necessitated by the bounds of our knowledge as such" (Schlick 1915, 154-155). As an example of an arbitrary element in theory that does not correspond to reality, Schlick cited the ether frame in Lorentzian electrodynamics (Schlick 1917, 60).

Einstein does not explicitly commend Schlick's defense of simplicity, but he also in no way objects in the course of a long correspondence during the late 1910s, wherein he strongly commends Schlick's general philosophical orientation and carefully records all points of disagreement (for more detail, see Howard 1984). Moreover the principle underlying Schlick's defense of simplicity, the idea that it is the non-arbitrary elements of our theories that represent the real, played a deep and enduring role in Einstein's philosophy of science.

UNIVOCALNESS IN THE THEORETICAL REPRESENTATION OF NATURE

In the physics and philosophy of science literature of the late nineteenth and early twentieth centuries, the principle according to which scientific theorizing should strive for a univocal representation of nature was widely and well known under the name that it was given in the title of a widely-cited essay by Joseph Petzoldt, "The Law of Univocalness" ["Das Gesetz der Eindeutigkeit"] (Petzoldt 1895). An indication that the map of philosophical positions was drawn then in a manner very different from today is to found in the fact that this principle found favor among both anti-metaphysical logical empiricists, such as Carnap, and neo-Kantians, such as Cassirer. It played a major role in debates over the ontology of general relativity and was an important part of the background to the development of the modern concept of categoricity in formal semantics (for more on the history, influence, and demise of the principle of univocalness, see Howard 1992 and 1996). One can find no more ardent and consistent champion of the principle than Einstein.

The principle of univocalness should not be mistaken for a denial of the underdetermination thesis. The latter asserts that a multiplicity of theories can equally well account for a given body of empirical evidence, perhaps even the infinity of all possible evidence in the extreme, Quinean version of the thesis. The principle of univocalness asserts (in a somewhat anachronistic formulation) that any one theory, even any one among a set of empirically equivalent theories, should provide a univocal representation of nature by determining for itself an isomorphic set of models. The unambiguous determination of theory choice by evidence is not the same thing as the univocal determination of a class of models by a theory.

The principle of univocalness played a central role in Einstein's struggles to formulate the general theory of relativity. When, in 1913, Einstein wrongly rejected a fully generally covariant theory of gravitation, he did so in part because he thought, wrongly, that generally covariant field equations failed the test of univocalness. More specifically, he reasoned wrongly that for a region of spacetime devoid of matter and energy-a "hole"-generally covariant field equations permit the construction of two different solutions, different in the sense that, in general, for spacetime points inside the hole, they assign different values of the metric tensor to one and the same point (for more on the history of this episode, see Stachel 1980 and Norton 1984). But Einstein's "hole argument" is wrong, and his own diagnosis of the error in 1915 rests again, ironically, on a deployment of the principle of univocalness.

What Einstein realized in 1915 was that, in 1913, he was wrongly assuming that a coordinate chart sufficed to fix the identity of spacetime manifold points. The application of a coordinate chart cannot suffice to individuate manifold points precisely because a coordinate chart is not an invariant labeling scheme, whereas univocalness in the representation of nature requires such invariance (see Howard and Norton 1993 and Howard 1999 for further discussion).

Here is how Einstein explained his change of perspective in a letter to Paul Ehrenfest of 26 December 1915, just a few weeks after the publication of the final, generally covariant formulation of the general theory of relativity:

In §12 of my work of last year, everything is correct (in the first three paragraphs) up to that which is printed with emphasis at the end of the third paragraph. From the fact that the two systems G(x) and G?(x), referred to the same reference system, satisfy the conditions of the grav. field, no contradiction follows with the univocalness of events. That which was apparently compelling in these reflections founders immediately, if one considers that

1. the reference system signifies nothing real
2. that the (simultaneous) realization of two different g-systems (or better, two different grav. fields) in the same region of the continuum is impossible according to the nature of the theory.

In place of §12, the following reflections must appear. The physically real in the universe of events (in contrast to that which is dependent upon the choice of a reference system) consists in spatiotemporal coincidences.* [Footnote *: and in nothing else!] Real are, e.g., the intersections of two different world lines, or the statement that they do not intersect. Those statements that refer to the physically real therefore do not founder on any univocal coordinate transformation. If two systems of the gμv (or in general the variables employed in the description of the world) are so created that one can obtain the second from the first through mere spacetime transformation, then they are completely equivalent. For they have all spatiotemporal point coincidences in common, i.e., everything that is observable.

These reflections show at the same time how natural the demand for general covariance is. (EA 9-363, ECP 8-173)

Einstein's new point of view, according to which the physically real consists exclusively in that which can be constructed on the basis of spacetime coincidences, spacetime points, for example, being regarded as intersections of world lines, is now known as the "point-coincidence argument." Spacetime coincidences play this privileged ontic role because they are invariant and, thus, univocally determined. Spacetime coordinates lack such invariance, a circumstance that Einstein thereafter repeatedly formulated as the claim that space and time "thereby lose the last vestige of physical reality" (see, for example, Einstein to Ehrenfest, 5 January 1916, EA 9-372, ECP 8-180). One telling measure of the philosophical importance of Einstein's new perspective on the ontology of spacetime is the fact that Schlick devoted his first book, Raum und Zeit in den gegenwärtigen Physik (1917), a book for which Einstein had high praise (see Howard 1984 and 1999), to an exploration of the philosophical implications of the claim that space and time have thereby lost the last vestige of physical reality. Mention has already been made of Schlick's defense of an underdetermination thesis based on his doctrine of truth as unambiguous coordination. That view is here developed at considerable length. But what most interested Einstein was Schlick's discussion of the reality concept. Schlick argued that Mach was wrong to regard only the elements of sensation as real. Spacetime events, individuated invariantly as spacetime coincidences, have as much or more right to be taken as real, precisely because of the univocal manner of their determination. Einstein wholeheartedly agreed, though he ventured the above-quoted suggestion that one should distinguish the two kinds of reality-that of the elements and that of the spacetime events-on the ground that if "two different peoples" pursued physics independently of one another they were fated to agree about the elements but would almost surely produce different theoretical constructions at the level of the spacetime event ontology. Note, again, that underdetermination is not a failure of univocalness. Different though they will be, each people's theoretical construction of an event ontology would be expected to be univocal.

Schlick, of course, went on to become the founder of the Vienna Circle, a leading figure in the development of logical empiricism, a champion of verificationism. That being so, an important question arises about Schlick's interpretation of Einstein on the univocal determination of spacetime events as spacetime coincidences. The question is this: Do such univocal coincidences play such a privileged role because of their reality or because of their observability. Clearly the former-the reality of that which is univocally determined-is important. But are univocal spacetime coincidences real because, thanks to their invariance, they are observable? Or is their observability consequent upon their invariant reality? Einstein, himself, repeatedly stressed the observable character of spacetime coincidences, as in the 26 December 1915 letter to Ehrenfest quoted above (for additional references and a fuller discussion, see Howard 1999).

Schlick, still a self-described realist in 1917, was clear about the relationship between observability and reality. He distinguished macroscopic coincidences in the field of our sense experience, to which he does accord a privileged and foundational epistemic status, from the microscopic point coincidences that define an ontology of spacetime manifold points. Mapping the former onto the latter is, for Schlick, an important part of the business of confirmation, but the reality of the spacetime manifold points is in no way consequent upon their observability. Indeed, how, strictly speaking, can one even talk of the observation of infinitesimal spacetime coincidences of the kind encountered in the intersection of two world lines? In fact, the order of implication goes the other way: Spacetime events individuated as spacetime coincidences are real because they are invariant, and such observability as they might possess is consequent upon their status as invariant bits of physical reality. For Einstein, and for Schlick in 1917, understanding the latter-physical reality-is the goal of physical theory.

REALISM AND SEPARABILITY

As we have seen, Schlick's Raum und Zeit in den gegenwärtigen Physik promoted a realistic interpretation of the ontology of general relativity. After reading the manuscript early in 1917, Einstein wrote to Schlick on 21 May that "the last section 'Relations to Philosophy' seems to me excellent" (EA 21-618, ECP 8-343), just the sort of praise one would expect from a fellow realist. Three years earlier, the Bonn mathematician, Eduard Study, had written another well-known, indeed very well-known defense of realism, Die realistische Weltansicht und die Lehre vom Raume (1914). Einstein read it in September of 1918. Much of it he liked, especially the droll style, as he said to Study in a letter of 17 September (EA 22-301, ECP 8-618). Pressed by Study to say more about the points where he disagreed, Einstein replied on 25 September in a rather surprising way:

I am supposed to explain to you my doubts? By laying stress on these it will appear that I want to pick holes in you everywhere. But things are not so bad, because I do not feel comfortable and at home in any of the "isms." It always seems to me as though such an ism were strong only so long as it nourishes itself on the weakness of it counter-ism; but if the latter is struck dead, and it is alone on an open field, then it also turns out to be unsteady on its feet. So, away we go!

"The physical world is real." That is supposed to be the fundamental hypothesis. What does "hypothesis" mean here? For me, a hypothesis is a statement, whose truth must be assumed for the moment, but whose meaning must be raised above all ambiguity. The above statement appears to me, however, to be, in itself, meaningless, as if one said: "The physical world is cock-a-doodle-doo." It appears to me that the "real" is an intrinsically empty, meaningless category (pigeon hole), whose monstrous importance lies only in the fact that I can do certain things in it and not certain others. This division is, to be sure, not an arbitrary one, but instead ....

I concede that the natural sciences concern the "real," but I am still not a realist. (EA 22-307, ECP-8-624)

Lest there be any doubt that Einstein has little sympathy for the other side, he adds:

The positivist or pragmatist is strong as long as he battles against the opinion that there [are] concepts that are anchored in the "A priori." When, in his enthusiasm, [he] forgets that all knowledge consists [in] concepts and judgments, then that is a weakness that lies not in the nature of things but in his personal disposition just as with the senseless battle against hypotheses, cf. the clear book by Duhem. In any case, the railing against atoms rests upon this weakness. Oh, how hard things are for man in this world; the path to originality leads through unreason (in the sciences), through ugliness (in the arts)-at least the path that many find passable. (EA 22-307, ECP-8-624)

What could Einstein mean by saying that he concedes that the natural sciences concern the "real," but that he is "still not a realist" and that the "real" in the statement, "the physical world is real," is an "intrinsically empty, meaningless category"?

The answer might be that realism, for Einstein, is not a philosophical doctrine about the interpretation of scientific theories or the semantics of theoretical terms. For Einstein, realism is a physical postulate, one of a most interesting kind, as he explained on 18 March 1948 in a long note at the end of the manuscript of Max Born's Waynflete Lectures, Natural Philosophy of Cause and Chance (1949), which Born had sent to Einstein for commentary:

I just want to explain what I mean when I say that we should try to hold on to physical reality. We are, to be sure, all of us aware of the situation regarding what will turn out to be the basic foundational concepts in physics: the point-mass or the particle is surely not among them; the field, in the Faraday - Maxwell sense, might be, but not with certainty. But that which we conceive as existing ('actual') should somehow be localized in time and space. That is, the real in one part of space, A, should (in theory) somehow 'exist' independently of that which is thought of as real in another part of space, B. If a physical system stretches over the parts of space A and B, then what is present in B should somehow have an existence independent of what is present in A. What is actually present in B should thus not depend upon the type of measurement carried out in the part of space, A; it should also be independent of whether or not, after all, a measurement is made in A.

If one adheres to this program, then one can hardly view the quantum-theoretical description as a complete representation of the physically real. If one attempts, nevertheless, so to view it, then one must assume that the physically real in B undergoes a sudden change because of a measurement in A. My physical instincts bristle at that suggestion.

However, if one renounces the assumption that what is present in different parts of space has an independent, real existence, then I do not at all see what physics is supposed to describe. For what is thought to by a 'system' is, after all, just conventional, and I do not see how one is supposed to divide up the world objectively so that one can make statements about the parts. (Born 1969, 223-224; my translation)

Realism is thus the thesis of spatial separability, the claim that spatial separation is a sufficient condition for the individuation of physical systems, and its assumption is here made into almost a necessary condition for the possibility of an intelligible science of physics.

The postulate of spatial separability as that which undergirds the ontic independence and, hence, individual identities of the systems that physics describes was an important part of Einstein's thinking about the foundations of physics since at least the time of his very first paper on the quantum hypothesis in 1905 (Einstein 1905a; for more detail on the early history of this idea in Einstein's thinking, see Howard 1990b). But the true significance of the separability principle emerged most clearly in 1935, when (as hinted in the just-quoted remark) Einstein made it one of the central premises of his argument for the incompleteness of quantum mechanics (see Howard 1985 and 1989). It is not so clearly deployed in the published version of the Einstein, Podolsky, Rosen paper (1935), but Einstein did not write that paper and did not like the way the argument appeared there. Separability is, however, an explicit premise in all of Einstein's later presentations of the argument for the incompleteness of quantum mechanics, both in correspondence and in print (see Howard 1985 for a detailed list of references).

In brief, the argument is this. Separability implies that spacelike separated systems have associated with them independent real states of affairs. A second postulate, locality, implies that the events in one region of spacetime cannot physically influence physical reality in a region of spacetime separated from the first by a spacelike interval. Consider now an experiment in which two systems, A and B, interact and separate, subsequent measurements on each corresponding to spacelike separated events. Separability implies that A and B have separate real physical states, and locality implies that the measurement performed on A cannot influence B's real physical state. But quantum mechanics ascribes different theoretical states, different wave functions, to B depending upon that parameter that is measured on A. Therefore, quantum mechanics ascribes different theoretical states to B, when B possesses, in fact, one real physical state. Hence quantum mechanics is incomplete.

One wants to ask many questions. First, what notion of completeness is being invoked here? It is not deductive completeness. It is closer in kind to what is termed "categoricity" in formal semantics, a categorical theory being one whose models are all isomorphic to one another. It is closer still to the principle discussed above-and cited as a precursor of the concept of categoricity-namely, the principle of univocalness, which we found doing such important work in Einstein's quest for a general theory of relativity, where it was the premise forcing the adoption of an invariant and thus univocal scheme for the individuation of spacetime manifold points.

The next question is why separability is viewed by Einstein as virtually an a priori necessary condition for the possibility of a science of physics. One reason is because a field theory like general relativity, which was Einstein's model for a future unified foundation for physics, is an extreme embodiment of the principle of separability: "Field theory has carried out this principle to the extreme, in that it localizes within infinitely small (four-dimensional) space-elements the elementary things existing independently of the one another that it takes as basic, as well as the elementary laws it postulates for them"

(Einstein 1948, 321-322). And a field theory like general relativity can do this because the infinitesimal metric interval-the careful way to think about separation in general relativistic spacetime-is invariant (hence univocally determined) under all continuous coordinate transformations.

Another reason why Einstein would be inclined to view separability as an a priori necessity is that, in thus invoking separability to ground individuation, Einstein places himself in a tradition of so viewing spatial separability with very strong Kantian roots (and, before Kant, Newtonian roots), a tradition in which spatial separability was known by the name that Arthur Schopenhauer famously gave to it, the principium individuationis (for a fuller discussion of this historical context, see Howard 1997).

A final question one wants to ask is: "What does any of this have to do with realism?" One might grant Einstein's point that a real ontology requires a principle of individuation without agreeing that separability provides the only conceivable such principle. Separability together with the invariance of the infinitesimal metric interval implies that, in a general relativistic spacetime, there are joints everywhere, meaning that we can carve up the universe in any way we choose and still have ontically independent parts. But quantum entanglement can be read as implying that this libertarian scheme of individuation does not work. Can quantum mechanics not be given a realistic interpretation? Many would say, "yes." Einstein said, "no."

6. The Principle Theories-Constructive Theories Distinction

There is much that is original in Einstein's philosophy of science as described thus far. At the very least, he rearranged the bits and pieces of doctrine that he learned from others-Kant, Mach, Duhem, Poincaré, Schlick, and others-in a strikingly novel way. But Einstein's most original contribution to twentieth-century philosophy of science lies elsewhere, in his distinction between what he termed "principle theories" and "constructive theories."

This idea first found its way into print in a brief 1919 article in the Times of London (Einstein 1919). A constructive theory, as the name implies, provides a constructive model for the phenomena of interest. An example would be kinetic theory. A principle theory consists of a set of individually well-confirmed, high-level empirical generalizations. Examples include the first and second laws of thermodynamics. Ultimate understanding requires a constructive theory, but often, says Einstein, progress in theory is impedea by premature attempts at developing constructive theories in the absence of sufficient constraints by means of which to narrow the range of possible of constructive. It is the function of principle theories to provide such constraint, and progress is often best achieved by focusing first on the establishment of such principles. According to Einstein, that is how he achieved his breakthrough with the theory of relativity, which, he says, is a principle theory, its two principles being the relativity principle and the light principle.

While the principle theories-constructive theories distinction first made its way into print in 1919, there is considerable evidence that it played an explicit role in Einstein's thinking much earlier. Nor was it only the relativity and light principles that served Einstein as constraints in his theorizing. Thus, he explicitly mentions also the Boltzmann principle, S = k log W, as another such:

> This equation connects thermodynamics with the molecular theory. It yields, as well, the statistical probabilities of the states of systems for which we are not in a position to construct a molecular-theoretical model. To that extent, Boltzmann's magnificent idea is of significance for theoretical physics ... because it provides a heuristic principle whose range extends beyond the domain of validity of molecular mechanics. (Einstein 1915, p. 262).

Einstein is here alluding the famous entropic analogy whereby, in his 1905 photon hypothesis paper, he reasoned from the fact that black body radiation in the Wien regime satisfied the Boltzmann principle to the conclusion that, in that regime, radiation behaved as if it consisted of mutually independent, corpuscle-like quanta of electromagnetic energy. The quantum hypothesis is a constructive model of radiation; the Boltzmann principle is the constraint that first suggested that model.

There are anticipations of the principle theories-constructive theories distinction in the nineteenth-century electrodynamics literature, James Clerk Maxwell, in particular, being a source from which Einstein might well have drawn (see Harman 1998). At the turn of the century, the "physics of principles" was a subject under wide discussion (see, for example, Poincaré 1904; for further discussion, see Giedymin 1982). But however extensive his borrowings (no explicit debt was ever acknowledged), in Einstein's hands the distinction becomes a methodological tool of impressive scope and fertility. What is puzzling, and even a bit sad, is that this most original methodological insight of Einstein's had comparatively little impact on later philosophy of science or practice in physics.

CONCLUSION: ALBERT EINSTEIN: PHILOSOPHER-PHYSICIST

Einstein's influence on twentieth-century philosophy of science is comparable to his influence on twentieth-century physics. What made that possible? One explanation looks to the institutional and disciplinary history of theoretical physics and the philosophy of science. Each was, in its own domain, a new mode of thought in the latter nineteenth century, and each finally began to secure for itself a solid institutional basis in the early twentieth century. In a curious way, the two movements helped one another. Philosophers of science helped to legitimate theoretical physics by locating the significant cognitive content of science in its theories. Theoretical physicists helped to legitimate the philosophy of science by providing for analysis a subject matter that was radically reshaping our understanding of nature and the place of humankind within it. In some cases the help was even more direct, as with the work of Einstein and Max Planck in the mid-1920s to create in the physics department at the University of Berlin a chair in the philosophy of science for Reichenbach (see Hecht and Hartmann 1982). And we should remember the example of the physicists Mach and Ludwig Boltzmann who were the first two occupants of the new chair for the philosophy of science at the University of Vienna at the turn of the century.

Another explanation looks to the education of young physicists in Einstein's day. Not only was Einstein's own youthful reading heavily focused on philosophy, more generally, and the philosophy of science, in particular (for an overview, see Einstein 1989, xxiv-xxv; see also Howard 1994b), in which respect he was not unlike other physicists of his generation, but also his university physics curriculum included a required course on "The Theory of Scientific Thought" (see Einstein 1987, Doc. 28). An obvious question is whether or not the early cultivation of a philosophical habit of mind made a difference in the way Einstein and his contemporaries approached physics. As indicated by his November 1944 letter to Robert Thorton quoted at the beginning of this article, Einstein thought that it did.

8

Systematic Epistemology

Introduction: Was Einstein an Epistemological "Opportunist"?

Late in 1944, Albert Einstein received a letter from Robert Thornton, a young African-American philosopher of science who had just finished his Ph.D. under Herbert Feigl at Minnesota and was beginning a new job teaching physics at the University of Puerto Rico, Mayaguez. He had written to solicit from Einstein a few supportive words on behalf of his efforts to introduce "as much of the philosophy of science as possible" into the modern physics course that he was to teach the following spring (Thornton to Einstein, 28 November 1944, EA 61-573). Here is what Einstein offered in reply:

I fully agree with you about the significance and educational value of methodology as well as history and philosophy of science. So many people today-and even professional scientists-seem to me like somebody who has seen thousands of trees but has never seen a forest. A knowledge of the historic and philosophical background gives that kind of independence from prejudices of his generation from which most scientists are suffering. This independence created by philosophical insight is-in my opinion-the mark of distinction between a mere artisan or specialist and a real seeker after truth. (Einstein to Thornton, 7 December 1944, EA 61-574)

That Einstein meant what he said about the relevance of philosophy to physics is evidenced by the fact that he had been saying more or less the same thing for decades. Thus, in a 1916 memorial note for Ernst Mach, a physicist and philosopher to whom Einstein owed a special debt, he wrote:

How does it happen that a properly endowed natural scientist comes to concern himself with epistemology? Is there no more valuable work in his specialty? I hear many of my colleagues saying, and I sense it from many more, that they feel this way. I cannot share this sentiment. When I think about the ablest students whom I have encountered in my teaching, that is, those who distinguish themselves by their independence of judgment and not merely their quick-wittedness, I can affirm that they had a vigorous interest in epistemology. They happily began discussions about the goals and methods of science, and they showed unequivocally, through their tenacity in defending their views, that the subject seemed important to them. Indeed, one should not be surprised at this. (Einstein 1916, 101)

How, exactly, does the philosophical habit of mind provide the physicist with such "independence of judgment"? Einstein goes on to explain:

Concepts that have proven useful in ordering things easily achieve such an authority over us that we forget their earthly origins and accept them as unalterable givens. Thus they come to be stamped as "necessities of thought," "a priori givens," etc. The path of scientific advance is often made impassable for a long time through such errors. For that reason, it is by no means an idle game if we become practiced in analyzing the long commonplace concepts and exhibiting those circumstances upon which their justification and usefulness depend, how they have grown up, individually, out of the givens of experience. By this means, their all-too-great authority will be broken. They will be removed if they cannot be properly legitimated, corrected if their correlation with given things be far too superfluous, replaced by others if a new system can be established that we prefer for whatever reason. (Einstein 1916, 102)

One is not surprised at Einstein's then citing Mach's critical analysis of the Newtonian conception of absolute space as a paradigm of what Mach, himself, termed the "historical-critical" method of philosophical analysis (Einstein 1916, 101, citing Ch. 2, §§ 6-7 of Mach's Mechanik, most likely the third edition, Mach 1897).

The place of philosophy in physics was a theme to which Einstein returned time and again, it being clearly an issue of deep importance to him. Sometimes he adopts a modest pose, as in this oft-quoted remark from his 1933 Spencer Lecture:

If you wish to learn from the theoretical physicist anything about the methods which he uses, I would give you the following piece of advice: Don't listen to his words, examine his achievements. For to the discoverer in that field, the constructions of his imagination appear so necessary and so natural that he is apt to treat them not as the creations of his thoughts but as given realities. (Einstein 1933, 5-6)

More typical, however, is the confident pose he struck three years later in "Physics and Reality":

It has often been said, and certainly not without justification, that the man of science is a poor philosopher. Why then should it not be the right thing for the physicist to let the philosopher do the philosophizing? Such might indeed be the right thing at a time when the physicist believes he has at his disposal a rigid system of fundamental concepts and fundamental laws which are so well established that waves of doubt can not reach them; but it can not be right at a time when the very foundations of physics itself have become problematic as they are now. At a time like the present, when experience forces us to seek a newer and more solid foundation, the physicist cannot simply surrender to the philosopher the critical contemplation of the theoretical foundations; for, he himself knows best, and feels more surely where the shoe pinches. In looking for a new foundation, he must try to make clear in his own mind just how far the concepts which he uses are justified, and are necessities. (Einstein 1936, 349)

What kind of philosophy might we expect from the philosopher-physicist? One thing that we should not expect from a physicist who takes the philosophical turn in order to help solve fundamental physical problems is a systematic philosophy:

The reciprocal relationship of epistemology and science is of noteworthy kind. They are dependent upon each other. Epistemology without contact with science becomes an empty scheme. Science without epistemology is-insofar as it is thinkable at all-primitive and muddled. However, no sooner has the epistemologist, who is seeking a clear system, fought his way through to such a system, than he is inclined to interpret the thought-content of science in the sense of his system and to reject whatever does not fit into his system. The scientist, however, cannot afford to carry his striving for epistemological systematic that far. He accepts gratefully the epistemological conceptual analysis; but the external conditions, which are set for him by the facts of experience, do not permit him to let himself be too much restricted in the construction of his conceptual world by the adherence to an epistemological system. He therefore must appear to the systematic epistemologist as a type of unscrupulous opportunist: he appears as realist insofar as he seeks to describe a world independent of the acts of perception; as idealist insofar as he looks upon the concepts and theories as free inventions of the human spirit (not logically derivable from what is empirically given); as positivist insofar as he considers his concepts and theories justified only to the extent to which they furnish a logical representation of relations among sensory experiences. He may even appear as Platonist or Pythagorean insofar as he considers the viewpoint of logical simplicity as an indispensable and effective tool of his research. (Einstein 1949, 683-684)

But what strikes the "systematic epistemologist" as mere opportunism might appear otherwise when viewed from the perspective of a physicist engaged, as Einstein himself put it, in "the critical contemplation of the theoretical foundations." The overarching goal of that critical contemplation was, for Einstein, the creation of a unified foundation for physics after the model of a field theory like general relativity. Einstein failed in his quest, but there was a consistency and constancy in the striving that informed as well the philosophy of science developing hand in hand with the scientific project.

Indeed, from early to late a few key ideas played the central, leading role in Einstein's philosophy of science, ideas about which Einstein evinced surprisingly little doubt even while achieving an ever deeper understanding of their implications. For the purposes of the following comparatively brief overview, we can confine our attention to just five topics:

- The underdetermination of theory choice by evidence.
- Simplicity and theory choice.
- Univocalness in the theoretical representation of nature.
- Realism and separability.
- The principle theories-constructive theories distinction.

In emphasizing the continuity and coherence in the development of Einstein's philosophy of science, I take issue with an account such as Gerald Holton's (1968), which claims to find a major philosophical break in the mid-1910s, in the form of a turn away from a sympathy for an anti-metaphysical positivism and toward a robust scientific realism. Holton sees this turn being driven by Einstein's alleged realization that general relativity, by contrast with special relativity, requires a realistic ontology. On my view, Einstein was never an ardent "Machian" positivist, and he was never a scientific realist, at least not in the sense acquired by the term "scientific realist" in later twentieth century philosophical discourse (see Howard 1993). Einstein expected scientific theories to have the proper empirical credentials, but he was no positivist; and he expected scientific theories to give an account of physical reality, but he was no scientific realist. Moreover, in both respects his views remained more or less the same from the beginning to the end of his career.

Why Einstein did not think himself a realist (he said so explicitly) is discussed below. Why he is not to be understood as a positivist deserves a word or two of further discussion here, if only because the belief that he was sympathetic to positivism, at least early in his life, is so widespread (for a fuller discussion, see Howard 1993).

That Einstein later repudiated positivism is beyond doubt. Many remarks from at least the early 1920s through the end of his life make this clear. In 1946 he explained what he took to be Mach's basic error:

He did not place in the correct light the essentially constructive and speculative nature of all thinking and more especially of scientific thinking; in consequence, he condemned theory precisely at those points where its constructive-speculative character comes to light unmistakably, such as in the kinetic theory of atoms. (Einstein 1946, 21)

Is Einstein here also criticizing his own youthful philosophical indiscretions? The very example that Einstein gives here makes any such interpretation highly implausible, because one of Einstein's main goals in his early work on Brownian motion (Einstein 1905b) was precisely to prove the reality of atoms, this in the face of the then famous skepticism of thinkers like Mach and Wilhelm Ostwald:

My principal aim in this was to find facts that would guarantee as much as possible the existence of atoms of definite size.... The agreement of these considerations with experience together with Planck's determination of the true molecular size from the law of radiation (for high temperatures) convinced the skeptics, who were quite numerous at that time (Ostwald, Mach), of the reality of atoms. (Einstein 1946, 45, 47)

Why, then, is the belief in Einstein's early sympathy for positivism so well entrenched?

The one piece of evidence standardly cited for a youthful flirtation with positivism is Einstein's critique of the notion of absolute distant simultaneity in his 1905 paper on special relativity (Einstein 1905c). Einstein speaks there of "observers," but in an epistemologically neutral way that can be replaced by talk of an inertial frame of reference. What really bothers Einstein about distant simultaneity is not that it is observationally inaccessible but that it involves a two-fold arbitrariness, one in the choice of an inertial frame of reference and one in the stipulation within a given frame of a convention regarding the ratio of the times required for a light signal to go from one stationary observer to another and back again. Likewise, Einstein faults classical Maxwellian electrodynamics for an asymmetry in the way it explains electromagnetic induction depending on whether it is the coil or the magnet that is assumed to be at rest. If the effect is the same-a current in the coil-why, asks Einstein, should there be two different explanations: an electrical field created in the vicinity of a moving magnet or an electromotive force induced in a conductor moving through a stationary magnetic field? To be sure, whether it is the coil or the magnet that is taken to be at rest makes no observable difference, but the problem, from Einstein's point of view, is the asymmetry in the two explanations. Even the young Einstein was no positivist.

First generation logical empiricists sought to legitimate their movement in part by claiming Einstein as a friend. They may be forgiven their putting a forced interpretation on arguments taken out of context. We can do better.

Einstein's philosophy of science is an original synthesis drawing upon many philosophical resources, from neo-Kantianism to Machian empiricism and Duhemian conventionalism. Other thinkers and movements, most notably the logical empiricists, drew upon the same resources. But Einstein put the pieces together in a manner importantly different from Moritz Schlick, Hans Reichenbach, and Rudolf Carnap, and he argued with them for decades about who was right (however much they obscured these differences in representing Einstein publicly as a friend of logical empiricism and scientific philosophy). Understanding how Einstein puts those pieces together therefore sheds light not only on the philosophical aspect of his own achievements in physics but also upon the larger history of the development of the philosophy of science in the twentieth century.

THE UNDERDETERMINATION OF THEORY CHOICE BY EVIDENCE: THE NATURE AND ROLE OF CONVENTIONS IN SCIENCE

Any philosophy of science must include an account of the relation between theory and evidence. Einstein learned about the historicity of scientific concepts from Mach. But his preferred way of modeling the logical relationship between theory and evidence was inspired mainly by his reading of Pierre Duhem's La Théorie physique: son objet et sa structure (Duhem 1906). Einstein probably first read Duhem, or at least learned the essentials of Duhem's philosophy of science around the fall of 1909, when, upon returning to Zurich from the patent office in Bern to take up his first academic appointment at the University of Zurich, he became the upstairs neighbor of his old friend and fellow Zurich physics student, Friedrich Adler. Just a few months before, Adler had published the German translation of La Théorie physique (Duhem 1908), and the philosophy of science became a frequent topic of conversation between the new neighbors, Adler and Einstein (see Howard 1990a).

Theoretical holism and the underdetermination of theory choice by empirical evidence are the central theses in Duhem's philosophy of science. His argument, in brief, is that at least in sciences like physics, where experiment is dense with sophisticated instrumentation whose employment itself requires theoretical interpretation, hypotheses are not tested in isolation but only as part of whole bodies of theory. It follows that when there is a conflict between theory and evidence, the fit can be restored in a multiplicity of different ways. No statement is immune to revision because of a presumed status as a definition or thanks to some other a priori warrant, and most any statement can be retained on pain of suitable adjustments elsewhere in the total body of theory. Hence, theory choice is underdetermined by evidence.

That Einstein's exposure to Duhem's philosophy of science soon left its mark is evident from lecture notes that Einstein prepared for a course on electricity and magnetism at the University of Zurich in the winter semester of 1910/11. Einstein asks how one can assign a definite electrical charge everywhere within a material body, if the interior of the body is not accessible to test particles. A "Machian" positivist would deem such direct empirical access necessary for meaningful talk of a charge distribution in the interior of a sold. Einstein argues otherwise:

We have seen how experience led to the introd. of the concept of the quantity of electricity. it was defined by means of the forces that small electrified bodies exert on each other. But now we extend the application of the concept to cases in which this definition cannot be applied directly as soon as we conceive the el. forces as forces exerted on electricity rather than on material particles. We set up a conceptual system the individual parts of which do not correspond directly to empirical facts. Only a certain totality of theoretical material corresponds again to a certain totality of experimental facts.

We find that such an el. continuum is always applicable only for the representation of el. states of affairs in the interior of ponderable bodies. Here too we define the vector of el. field strength as the vector of the mech. force exerted on the unit of pos. electr. quantity inside a body. But the force so defined is no longer directly accessible to exp. It is one part of a theoretical construction that can be correct or false, i.e., consistent or not consistent with experience, only as a whole. (EA 3-007, ECP 3-11, 325) One can hardly ask for a better summary of Duhem's point of view in application to a specific physical theory.

Explicit citations of Duhem by Einstein are rare (for details, see Howard 1990a). But explicit invocations of a holist picture of the structure and empirical interpretation of theories are not hard to find. An especially interesting example is found in a review that Einstein wrote in 1924 of Alfred Elsbach's Kant und Einstein (1924), one of the flood of books and articles then trying to reconcile the Kantian doctrine of the a priori Euclidean character of space with general relativity's postulate of variable spacetime curvature. Having asserted that relativity theory is incompatible with Kant's doctrine of the a priori, Einstein explains why, more generally, he is not sympathetic with Kant:

This does not, at first, preclude one's holding at least to the Kantian problematic, as, e.g., Cassirer has done. I am even of the opinion that this standpoint can be rigorously refuted by no development of natural science. For one will always be able to say that critical philosophers have until now erred in the establishment of the a priori elements, and one will always be able to establish a system of a priori elements that does not contradict a given physical system. Let me briefly indicate why I do not find this standpoint natural. A physical theory consists of the parts (elements) A, B, C, D, that together constitute a logical whole which correctly connects the pertinent experiments (sense experiences). Then it tends to be the case that the aggregate of fewer than all four elements, e.g., A, B, D, without C, no longer says anything about these experiences, and just as well A, B, C without D. One is then free to regard the aggregate of three of these elements, e.g., A, B, C as a priori, and only D as empirically conditioned. But what remains unsatisfactory in this is always the arbitrariness in the choice of those elements that one designates as a priori, entirely apart from the fact that the theory could one day be replaced by another that replaces certain of these elements (or all four) by others. (Einstein 1924, 1688-1689)

Einstein's point seems to be that while one can always choose to designate selected elements as a priori and, hence, non-empirical, no principle determines which elements can be so designated, and our ability thus to designate them derives from the fact that it is only the totality of the elements that possesses empirical content.

Much the same point could be made, and was made by Duhem himself (see Duhem 1906, part 2, ch. 6, sects. 8 and 9), against those who would insulate certain statements against empirical refutation by claiming for them the status of conventional definitions. Edouard Le Roy (1901) had argued thus about the law of free fall. It could not be refuted by experiment because it functioned as a definition of "free fall." And Henri Poincaré (1901) said much the same about the principles of mechanics more generally. As Einstein answered the neo-Kantians, so Duhem answered this species of conventionalist: Yes, experiment cannot refute, say, the law of free fall by itself, but only because it is part of a larger theoretical whole that has empirical content only as a whole, and various other elements of that whole could as well be said to be, alone, immune to refutation.

That Einstein should deploy against the neo-Kantians in the early 1920s the argument that Duhem used against the conventionalism of Poincaré and Le Roy is interesting from the point of view of Einstein's relationships with those who were leading the development of logical empiricism and scientific philosophy in the 1920s, especially Schlick and Reichenbach. Einstein shared with Schlick and Reichenbach the goal of crafting a new form of empiricism that would be adequate to the task of defending general relativity against neo-Kantian critiques (see Schlick 1917 and 1921, and Reichenbach 1920, 1924, and 1928; for more detail, see Howard 1994a). But while they all agreed that what Kant regarded as the a priori element in scientific cognition was better understood as a conventional moment in science, they were growing to disagree dramatically over the nature and place of conventions in science. With the help of new logical tools and a more sophisticated verificationist semantics, Schlick and Reichenbach were refining Poincaré's idea of conventional definitional elements in science into the classic logical empiricist view that the moment of convention was restricted to conventional coordinating definitions that endow individual primitive terms and, by extension, the individual synthetic propositions constructed out of them with empirical content. This view worked well as an answer to the neo-Kantian, for it implied that once one fixed one's coordinating definition-as with a conventional choice of a standard measuring rod coordinated with the geometer's concept of a "rigid body"-the question of the curvature of space had an empirically determinate answer. But unless the division is wholly arbitrary, parsing theories thus into coordinating definitions and empirical statements assumes a principled difference in kind between the two categories of statements along the lines of an analytic-synthetic distinction. As had been the case with Duhem before him, the assumption of such a principled difference in kind did not comport well with the holism about theories that Einstein had learned from Duhem.

It was this argument over the nature and place of conventions in science that underlay Einstein's gradual philosophical estrangement from Schlick and Reichenbach in the 1920s. Serious in its own right, the argument over conventions was entangled with two other issues as well, namely, realism and Einstein's famous view of theories as the "free creations of the human spirit" (see, for example, Einstein 1921). In both instances what troubled Einstein was that a verificationist semantics made the link between theory and experience too strong, leaving too small a role for theory, itself, and the creative theorizing that produces it.

If theory choice is empirically determinate, especially if theoretical concepts are explicitly constructed from empirical primitives, as in Carnap's program in the Aufbau (Carnap 1928), then it is hard to see how theory gives us a story about anything other than experience. As noted, Einstein was not what we would today call a scientific realist, but he still believed that there was content in theory beyond mere empirical content. He believed that theoretical science gave us a window on nature itself, even if, in principle, there will be no one uniquely correct story at the level of deep ontology (see below, section 5). And if the only choice in theory choice is one among conventional coordinating definitions, then that is no choice at all, a point stressed by Reichenbach, especially, as an important positive implication of his position. Reichenbach argued that if empirical content is the only content, then empirically equivalent theories have the same content, the difference resulting from their different choices of coordinating definitions being like in kind to the difference between "es regnet" and "il pleut," or the difference between expressing the result of a measurement in English or metric units, just two different ways of saying the same thing. But then, Einstein would ask, where is there any role for the creative intelligence of the theoretical physicist if there is no room for genuine choice in science, if experience somehow dictates theory construction?

The argument over the nature and role of conventions in science continued to the very end of Einstein's life, reaching its highest level of sophistication in the exchange between Reichenbach and Einstein the Library of Living Philosopher's volume, Albert Einstein: Philosopher-Physicist(Schilpp 1949). The question is, again, whether the choice of a geometry is empirical, conventional, or a priori. In his contribution, Reichenbach reasserted his old view that once an appropriate coordinating definition is established, equating some "practically rigid rod" with the geometer's "rigid body," then the geometry of physical space is wholly determined by empirical evidence:

The choice of a geometry is arbitrary only so long as no definition of congruence is specified. Once this definition is set up, it becomes an empirical question which geometry holds for physical space.... The conventionalist overlooks the fact that only the incomplete statement of a geometry, in which a reference to the definition of congruence is omitted, is arbitrary. (Reichenbach 1949, 297)

Einstein's clever reply includes a dialogue between two characters, "Reichenbach" and "Poincaré," in which "Reichenbach" concedes to "Poincaré" that there are no perfectly rigid bodies in nature and that physics must be used to correct for such things as thermal deformations, from which it follows that what we actually test is geometry plus physics, not geometry alone. Here an "anonymous non-positivist" takes "Poincaré's" place, out of respect, says Einstein, "for Poincaré's superiority as thinker and author" (Einstein 1949, 677), but also, perhaps, because he realized that the point of view that follows was more Duhem than Poincaré. The "non-positivist" then argues that one's granting that geometry and physics are tested together contravenes the positivist identification of meaning with verifiability:

Non-Positivist: If, under the stated circumstances, you hold distance to be a legitimate concept, how then is it with your basic principle (meaning = verifiability)? Must you not come to the point where you deny the meaning of geometrical statements and concede meaning only to the completely developed theory of relativity (which still does not exist at all as a finished product)? Must you not grant that no "meaning" whatsoever, in your sense, belongs to the individual concepts and statements of a physical theory, such meaning belonging instead to the whole system insofar as it makes "intelligible" what is given in experience? Why do the individual concepts that occur in a theory require any separate justification after all, if they are indispensable only within the framework of the logical structure of the theory, and if it is the theory as a whole that stands the test? (Einstein 1949, 678).

Two years before the Quine's publication of "Two Dogmas of Empiricism" (1951), Einstein here makes explicit the semantic implications of a thoroughgoing holism.

If theory choice is empirically underdetermined, then an obvious question is why we are so little aware of the underdetermination in the day-to-day conduct of science. In a 1918 address celebrating Max Planck's sixtieth birthday, Einstein approached this question via a distinction between practice and principle:

The supreme task of the physicist is ... the search for those most general, elementary laws from which the world picture is to be obtained through pure deduction. No logical path leads to these elementary laws; it is instead just the intuition that rests on an empathic understanding of experience. In this state of methodological uncertainty one can think that arbitrarily many, in themselves equally justified systems of theoretical principles were possible; and this opinion is, in principle, certainly correct. But the development of physics has shown that of all the conceivable theoretical constructions a single one has, at any given time, proved itself unconditionally superior to all others. No one who has really gone deeply into the subject will deny that, in practice, the world of perceptions determines the theoretical system unambiguously, even though no logical path leads from the perceptions to the basic principles of the theory. (Einstein 1918, 31; my translation)

But why is theory choice, in practice, seemingly empirically determined? Einstein hinted at an answer the year before in a letter to Schlick, where he commended Schlick's argument that the deep elements of a theoretical ontology have as much claim to the status of the real as do Mach's elements of sensation (Schlick 1917), but suggested that we are nonetheless speaking of two different kinds of reality. How do they differ?

It appears to me that the word "real" is taken in different senses, according to whether impressions or events, that is to say, states of affairs in the physical sense, are spoken of.

If two different peoples pursue physics independently of one another, they will create systems that certainly agree as regards the impressions ("elements" in Mach's sense). The mental constructions that the two devise for connecting these "elements" can be vastly different. And the two constructions need not agree as regards the "events"; for these surely belong to the conceptual constructions. Certainly on the "elements," but not the "events," are real in the sense of being "given unavoidably in experience."

But if we designate as "real" that which we arrange in the space-time-schema, as you have done in the theory of knowledge, then without doubt the "events," above all, are real.... I would like to recommend a clean conceptual distinction here. (Einstein to Schlick, 21 May 1917, EA 21-618, ECP 8-343)

Why, in practice, are physicists unaware of underdetermination? It is because ours is not the situation of "two different peoples pursu[ing] physics independently of one another." Though Einstein does not say it explicitly, the implication seems to be that apparent determination in theory choice is mainly a consequence of our all being similarly socialized as we become members of a common scientific community. Part of what it means to be a member of a sucha community is that we have been taught to make our theoretical choices in accord with criteria or values that we hold in common.

SIMPLICITY AND THEORY CHOICE

For Einstein, as for many others, simplicity is the criterion that mainly steers theory choice in domains where experiment and observation no longer provide an unambiguous guide. This, too, is a theme sounded early and late in Einstein's philosophical reflections (for more detail, see Howard 1998, Norton 2000, and van Dongen 2002). For example, the just-quoted remark from 1918 about the apparent determination of theory choice in practice, contrasted with in-principle underdetermination continues:

Furthermore this conceptual system that is univocally coordinated with the world of experience is reducible to a few basic laws from which the whole system can be developed logically. With every new important advance the researcher here sees his expectations surpassed, in that those basic laws are more and more simplified under the press of experience. With astonishment he sees apparent chaos resolved into a sublime order that is to be attributed not to the rule of the individual mind, but to the constitution of the world of experience; this is what Leibniz so happily characterized as "pre-established harmony." Physicists strenuously reproach many epistemologists for their insufficient appreciation of this circumstance. Herein, it seems to me, lie the roots of the controversy carried on some years ago between Mach and Planck. (Einstein 1918, p. 31)

There is more than a little autobiography here, for as Einstein stressed repeatedly in later years, he understood the success of his own quest for a general theory of relativity as a result of his seeking the simplest set of field equations satisfying a given set of constraints.

Always a leitmotif, Einstein's celebration of simplicity as a guide to theory choice comes clearly to the fore in the early 1930s. Why then? The reason might well be that his faith in simplicity had been vindicated when, seemingly with a sigh of relief, he found that he could drop from the gravitational field equations the cosmological constant that he had introduced in 1917 for the purpose of blocking non-static solutions, for the introduction of the cosmological constant in the first place had represented to him "a considerable renunciation of the logical simplicity of the theory" (Einstein 1949, 684-685). That his faith in simplicity was reaffirmed is clear. Witness what he wrote in his 1933 Herbert Spencer lecture:

If, then, it is true that the axiomatic foundation of theoretical physics cannot be extracted from experience but must be freely invented, may we ever hope to find the right way? Furthermore, does this right way exist anywhere other than in our illusions? May we hope to be guided safely by experience at all, if there exist theories (such as classical mechanics) which to a large extent do justice to experience, without comprehending the matter in a deep way?

To these questions, I answer with complete confidence, that, in my opinion, the right way exists, and that we are capable of finding it. Our experience hitherto justifies us in trusting that nature is the realization of the simplest that is mathematically conceivable. I am convinced that purely mathematical construction enables us to find those concepts and those lawlike connections between them that provide the key to the understanding of natural phenomena. Useful mathematical concepts may well be suggested by experience, but in no way can they be derived from it. Experience naturally remains the sole criterion of the usefulness of a mathematical construction for physics. But the actual creative principle lies in mathematics. Thus, in a certain sense, I take it to be true that pure thought can grasp the real, as the ancients had dreamed. (Einstein 1933, p. 183; my translation)

Another consideration reinforcing Einstein's conviction that the theoretical physicist must trust simplicity is that physics is moving steadily into domains ever further removed from direct contact with observation and experiment. Before the 1960s, general relativity, itself, rested on a famously thin empirical footing, and empirical evidence provided even less of a guide in Einstein's search for a unified field theory. One year after the Herbert Spencer lecture, at a time when he was immersed in work on unified field theory, Einstein wrote:

The theory of relativity is a beautiful example of the basic character of the modern development of theory. That is to say, the hypotheses from which one starts become ever more abstract and more remote from experience. But in return one comes closer to the preeminent goal of science, that of encompassing a maximum of empirical contents through logical deduction with a minimum of hypotheses or axioms. The intellectual path from the axioms to the empirical contents or to the testable consequences becomes, thereby, ever longer and more subtle. The theoretician is forced, ever more, to allow himself to be directed by purely mathematical, formal points of view in the search for theories, because the physical experience of the experimenter is not capable of leading us up to the regions of the highest abstraction. Tentative deduction takes the place of the predominantly inductive methods appropriate to the youthful state of science. Such a theoretical structure must be quite thoroughly elaborated in order for it to lead to consequences that can be compared with experience. It is certainly the case that here, as well, the empirical fact is the all-powerful judge. But its judgment can be handed down only on the basis of great and difficult intellectual effort that first bridges the wide space between the axioms and the testable consequences. The theorist must accomplish this Herculean task with the clear understanding that this effort may only be destined to prepare the way for a death sentence for his theory. One should not reproach the theorist who undertakes such a task by calling him a fantast; instead, one must allow him his fantasizing, since for him there is no other way to his goal whatsoever. Indeed, it is no planless fantasizing, but rather a search for the logically simplest possibilities and their consequences. (Einstein 1954, 238-239; my translation)

What warrant is there for thus trusting in simplicity? At best one can do a kind of meta-induction. That "the totality of all sensory experience can be 'comprehended' on the basis of a conceptual system built on premises of great simplicity" will be derided by skeptics as a "miracle creed," but, Einstein adds, "it is a miracle creed which has been borne out to an amazing extent by the development of science" (Einstein 1950, p. 342).

But for all that Einstein's faith in simplicity was strong, he despaired of giving a precise, formal characterization of how we assess the simplicity of a theory. In 1946 he wrote about the perspective of simplicity (here termed the "inner perfection" of a theory):

This point of view, whose exact formulation meets with great difficulties, has played an important role in the selection and evaluation of theories from time immemorial. The problem here is not simply one of a kind of enumeration of the logically independent premises (if anything like this were at all possible without ambiguity), but one of a kind of reciprocal weighing of incommensurable qualities.... I shall not attempt to excuse the lack of precision of [these] assertions ... on the grounds of insufficient space at my disposal; I must confess herewith that I cannot at this point, and perhaps not at all, replace these hints by more precise definitions. I believe, however, that a sharper formulation would be possible. In any case it turns out that among the "oracles" there usually is agreement in judging the "inner perfection" of the theories and even more so concerning the degree of "external confirmation." (Einstein 1946, pp. 21, 23).

As in 1918, so in 1946 and beyond, Einstein continues to be impressed that the "oracles," presumably the leaders of the relevant scientific community, tend to agree in their judgments of simplicity. That is why, in practice, simplicity seems to determine theory choice univocally.

Experience justifies our trusting that nature is the realization of the simplest that is mathematically conceivable. Can we say anything more about why this might be so? A hint is provided by Einstein's enthusiastically positive response to Schlick's first essay on the philosophical significance of relativity (Schlick 1915). At this early stage in his philosophical career, Schlick regarded himself as a realist and defended a version of the underdetermination thesis grounded in his view of truth as the unambiguous many-to-one coordination of propositions to facts (Schlick 1910). Theories being sets of propositions, several theories could likewise be unambiguously coordinated with a given set of facts and thus count as equally true representations of those facts. When he took up the question of simplicity, Schlick derided those who would justify simplicity as a criterion of theory choice by arguing that we should choose simple theories because nature itself is simple. As Schlick rightly pointed out, this is a circular argument, for our only cognitive access to nature is via our theories (note that Einstein argues not that nature, itself, is simple, but that nature is a realization of simple theoretical constructions, a crucial difference). Schlick similarly derides a defense grounded in considerations of mental economy, which he terms "intellectual indolence." Why then choose simple theories? Schlick's answer is that "the greater simplicity of a theory depends on its containing fewer arbitrary elements." Why is it better to choose theories with fewer "superfluous," arbitrary elements? Because only the non-arbitrary elements are likely to correspond to reality, so in choosing the simpler theory "we are then sure of diverging from reality at least no further than is necessitated by the bounds of our knowledge as such" (Schlick 1915, 154-155). As an example of an arbitrary element in theory that does not correspond to reality, Schlick cited the ether frame in Lorentzian electrodynamics (Schlick 1917, 60).

Einstein does not explicitly commend Schlick's defense of simplicity, but he also in no way objects in the course of a long correspondence during the late 1910s, wherein he strongly commends Schlick's general philosophical orientation and carefully records all points of disagreement (for more detail, see Howard 1984). Moreover the principle underlying Schlick's defense of simplicity, the idea that it is the non-arbitrary elements of our theories that represent the real, played a deep and enduring role in Einstein's philosophy of science.

UNIVOCALNESS IN THE THEORETICAL REPRESENTATION OF NATURE

In the physics and philosophy of science literature of the late nineteenth and early twentieth centuries, the principle according to which scientific theorizing should strive for a univocal representation of nature was widely and well known under the name that it was given in the title of a widely-cited essay by Joseph Petzoldt, "The Law of Univocalness" ["Das Gesetz der Eindeutigkeit"] (Petzoldt 1895). An indication that the map of philosophical positions was drawn then in a manner very different from today is to found in the fact that this principle found favor among both anti-metaphysical logical empiricists, such as Carnap, and neo-Kantians, such as Cassirer. It played a major role in debates over the ontology of general relativity and was an important part of the background to the development of the modern concept of categoricity in formal semantics (for more on the history, influence, and demise of the principle of univocalness, see Howard 1992 and 1996). One can find no more ardent and consistent champion of the principle than Einstein.

The principle of univocalness should not be mistaken for a denial of the underdetermination thesis. The latter asserts that a multiplicity of theories can equally well account for a given body of empirical evidence, perhaps even the infinity of all possible evidence in the extreme, Quinean version of the thesis. The principle of univocalness asserts (in a somewhat anachronistic formulation) that any one theory, even any one among a set of empirically equivalent theories, should provide a univocal representation of nature by determining for itself an isomorphic set of models. The unambiguous determination of theory choice by evidence is not the same thing as the univocal determination of a class of models by a theory.

The principle of univocalness played a central role in Einstein's struggles to formulate the general theory of relativity. When, in 1913, Einstein wrongly rejected a fully generally covariant theory of gravitation, he did so in part because he thought, wrongly, that generally covariant field equations failed the test of univocalness. More specifically, he reasoned wrongly that for a region of spacetime devoid of matter and energy-a "hole"-generally covariant field equations permit the construction of two different solutions, different in the sense that, in general, for spacetime points inside the hole, they assign different values of the metric tensor to one and the same point (for more on the history of this episode, see Stachel 1980 and Norton 1984). But Einstein's "hole argument" is wrong, and his own diagnosis of the error in 1915 rests again, ironically, on a deployment of the principle of univocalness. What Einstein realized in 1915 was that, in 1913, he was wrongly assuming that a coordinate chart sufficed to fix the identity of spacetime manifold points. The application of a coordinate chart cannot suffice to individuate manifold points precisely because a coordinate chart is not an invariant labeling scheme, whereas univocalness in the representation of nature requires such invariance (see Howard and Norton 1993 and Howard 1999 for further discussion).

Here is how Einstein explained his change of perspective in a letter to Paul Ehrenfest of 26 December 1915, just a few weeks after the publication of the final, generally covariant formulation of the general theory of relativity:

In §12 of my work of last year, everything is correct (in the first three paragraphs) up to that which is printed with emphasis at the end of the third paragraph. From the fact that the two systems G(x) and G?(x), referred to the same reference system, satisfy the conditions of the grav. field, no contradiction follows with the univocalness of events. That which was apparently compelling in these reflections founders immediately, if one considers that

THE REFERENCE SYSTEM SIGNIFIES NOTHING REAL

that the (simultaneous) realization of two different g-systems (or better, two different grav. fields) in the same region of the continuum is impossible according to the nature of the theory.

In place of §12, the following reflections must appear. The physically real in the universe of events (in contrast to that which is dependent upon the choice of a reference system) consists in spatiotemporal coincidences.* [Footnote *: and in nothing else!] Real are, e.g., the intersections of two different world lines, or the statement that they do not intersect. Those statements that refer to the physically real therefore do not founder on any univocal coordinate transformation. If two systems of the $g\mu\nu$ (or in general the variables employed in the description of the world) are so created that one can obtain the second from the first through mere spacetime transformation, then they are completely equivalent. For they have all spatiotemporal point coincidences in common, i.e., everything that is observable.

These reflections show at the same time how natural the demand for general covariance is. (EA 9-363, ECP 8-173)

Einstein's new point of view, according to which the physically real consists exclusively in that which can be constructed on the basis of spacetime coincidences, spacetime points, for example, being regarded as intersections of world lines, is now known as the "point-coincidence argument." Spacetime coincidences play this privileged ontic role because they are invariant and, thus, univocally determined. Spacetime coordinates lack such invariance, a circumstance that Einstein thereafter repeatedly formulated as the claim that space and time "thereby lose the last vestige of physical reality" (see, for example, Einstein to Ehrenfest, 5 January 1916, EA 9-372, ECP 8-180).

One telling measure of the philosophical importance of Einstein's new perspective on the ontology of spacetime is the fact that Schlick devoted his first book, Raum und Zeit in den gegenwärtigen Physik (1917), a book for which Einstein had high praise (see Howard 1984 and 1999), to an exploration of the philosophical implications of the claim that space and time have thereby lost the last vestige of physical reality. Mention has already been made of Schlick's defense of an underdetermination thesis based on his doctrine of truth as unambiguous coordination. That view is here developed at considerable length. But what most interested Einstein was Schlick's discussion of the reality concept. Schlick argued that Mach was wrong to regard only the elements of sensation as real. Spacetime events, individuated invariantly as spacetime coincidences, have as much or more right to be taken as real, precisely because of the univocal manner of their determination. Einstein wholeheartedly agreed, though he ventured the above-quoted suggestion that one should distinguish the two kinds of reality-that of the elements and that of the spacetime events-on the ground that if "two different peoples" pursued physics independently of one another they were fated to agree about the elements but would almost surely produce different theoretical constructions at the level of the spacetime event ontology. Note, again, that underdetermination is not a failure of univocalness. Different though they will be, each people's theoretical construction of an event ontology would be expected to be univocal.

Schlick, of course, went on to become the founder of the Vienna Circle, a leading figure in the development of logical empiricism, a champion of verificationism. That being so, an important question arises about Schlick's interpretation of Einstein on the univocal determination of spacetime events as spacetime coincidences. The question is this: Do such univocal coincidences play such a privileged role because of their reality or because of their observability. Clearly the former-the reality of that which is univocally determined-is important. But are univocal spacetime coincidences real because, thanks to their invariance, they are observable? Or is their observability consequent upon their invariant reality? Einstein, himself, repeatedly stressed the observable character of spacetime coincidences, as in the 26 December 1915 letter to Ehrenfest quoted above (for additional references and a fuller discussion, see Howard 1999).

Schlick, still a self-described realist in 1917, was clear about the relationship between observability and reality. He distinguished macroscopic coincidences in the field of our sense experience, to which he does accord a privileged and foundational epistemic status, fı ɔm the microscopic point coincidences that define an ontology of spacetime manifold points. Mapping the former onto the latter is, for Schlick, an important part of the business of confirmation, but the reality of the spacetime manifold points is in no way consequent upon their observability. Indeed, how, strictly speaking, can one even talk of the observation of infinitesimal spacetime coincidences of the kind encountered in the intersection of two world lines? In fact, the order of implication goes the other way: Spacetime events individuated as spacetime coincidences are real because they are invariant, and such observability as they might possess is consequent upon their status as invariant bits of physical reality. For Einstein, and for Schlick in 1917, understanding the latter-physical reality-is the goal of physical theory.

REALISM AND SEPARABILITY

As we have seen, Schlick's Raum und Zeit in den gegenwärtigen Physik promoted a realistic interpretation of the ontology of general relativity. After reading the manuscript early in 1917, Einstein wrote to Schlick on 21 May that "the last section 'Relations to Philosophy' seems to me excellent" (EA 21-618, ECP 8-343), just the sort of praise one would expect from a fellow realist. Three years earlier, the Bonn mathematician, Eduard Study, had written another well-known, indeed very well-known defense of realism, Die realistische Weltansicht und die Lehre vom Raume (1914). Einstein read it in September of 1918. Much of it he liked, especially the droll style, as he said to Study in a letter of 17 September (EA 22-301, ECP 8-618). Pressed by Study to say more about the points where he disagreed, Einstein replied on 25 September in a rather surprising way:

The positivist or pragmatist is strong as long as he battles against the opinion that there [are] concepts that are anchored in the "A priori." When, in his enthusiasm, [he] forgets that all knowledge consists [in] concepts and judgments, then that is a weakness that lies not in the nature of things but in his personal disposition just as with the senseless battle against hypotheses, cf. the clear book by Duhem. In any case, the railing against atoms rests upon this weakness. Oh, how hard things are for man in this world; the path to originality leads through unreason (in the sciences), through ugliness (in the arts)-at least the path that many find passable. (EA 22-307, ECP-8-624) What could Einstein mean by saying that he concedes that the natural sciences concern the "real," but that he is "still not a realist" and that the "real" in the statement, "the physical world is real," is an "intrinsically empty, meaninglessv category"? The answer might be that realism, for Einstein, is not a philosophical doctrine about the interpretation of scientific theories or the semantics of theoretical terms. For Einstein, realism is a physical postulate, one of a most interesting kind, as he explained on 18 March 1948 in a long note at the end of the manuscript of Max Born's Waynflete Lectures, Natural Philosophy of Cause and Chance (1949), which Born had sent to Einstein for commentary:

Bibliography

- G.A.G. Bennet (1974). Electricity and Modern Physics (2nd ed.). Edward Arnold (UK). ISBN 0-7131-2459-8.
- Browne, Michael (2008). Physics for Engineering and Science, (2nd ed.). McGraw-Hill/Schaum. ISBN 978-0-07-161399-6.
- Dibner, Bern (2012). Oersted and the discovery of electromagnetism. Literary Licensing, LLC. ISBN 978-1-258-33555-7.
- Durney, Carl H.; Johnson, Curtis C. (1969). Introduction to modern electromagnetics. McGraw-Hill. ISBN 0-07-018388-0.
- Feynman, Richard P. (1970). The Feynman Lectures on Physics Vol II. Addison Wesley Longman. ISBN 978-0-201-02115-8.
- Fleisch, Daniel (2008). A Student's Guide to Maxwell's Equations. Cambridge, UK: Cambridge University Press. ISBN 978-0-521-70147-1.
- I.S. Grant; W.R. Phillips; Manchester Physics (2008). Electromagnetism (2nd ed.). John Wiley & Sons. ISBN 978-0-471-92712-9.
- Griffiths, David J. (1998). Introduction to Electrodynamics (3rd ed.). Prentice Hall. ISBN 0-13-805326-X.
- Jackson, John D. (1998). Classical Electrodynamics (3rd ed.). Wiley. ISBN 0-471-30932-X.
- Moliton, André (2007). Basic electromagnetism and materials. 430 pages. New York City: Springer-Verlag New York, LLC. ISBN 978-0-387-30284-3.
- Purcell, Edward M. (1985). Electricity and Magnetism Berkeley, Physics Course Volume 2 (2nd ed.). McGraw-Hill. ISBN 0-07-004908-4.
- Purcell, Edward M and Morin, David. (2013). Electricity and Magnetism, 820p, (3rd ed.). Cambridge University Press, New York. ISBN 978-1-107-01402-2.
- Rao, Nannapaneni N. (1994). Elements of engineering electromagnetics (4th ed.). Prentice Hall. ISBN 0-13-948746-8.
- Rothwell, Edward J.; Cloud, Michael J. (2001). Electromagnetics. CRC Press. ISBN 0-8493-1397-X.

- Tipler, Paul (1998). Physics for Scientists and Engineers: Vol. 2: Light, Electricity and Magnetism (4th ed.). W. H. Freeman. ISBN 1-57259-492-6.
- Wangsness, Roald K.; Cloud, Michael J. (1986). Electromagnetic Fields (2nd Edition). Wiley. ISBN 0-471-81186-6.
- Poyser, A. W. (1892). Magnetism and Electricity: A Manual for Students in Advanced Classes. London and New York: Longmans, Green, & Co. p. 285.
- Giancoli, Douglas C. (1998). Physics: Principles with Applications (Fifth ed.). pp. 623-624.
- Ulaby, Fawwaz (2007). Fundamentals of applied electromagnetics (5th ed.). Pearson:Prentice Hall. p. 255. ISBN 0-13-241326-4.
- Joseph Henry". Distinguished Members Gallery, National Academy of Sciences. Archived from the original on 2013-12-13. Retrieved 2006-11-30.
- A Brief History of Electromagnetism" (PDF).
- Electromagnetism". Smithsonian Institution Archives.
- Michael Faraday, by L. Pearce Williams, p. 182-3
- Michael Faraday, by L. Pearce Williams, p. 191-5
- Michael Faraday, by L. Pearce Williams, p. 510
- Maxwell, James Clerk (1904), A Treatise on Electricity and Magnetism, Vol. II, Third Edition. Oxford University Press, pp. 178-9 and 189.
- Archives Biographies: Michael Faraday", The Institution of Engineering and Technology.
- Good, R. H. (1999). Classical Electromagnetism. Saunders College Publishing. p. 107. ISBN 0-03-022353-9.
- Feynman, R. P.; Leighton, R. B.; Sands, M. L. (2006). The Feynman Lectures on Physics, Volume 2. Pearson/Addison-Wesley. pp. 17-2. ISBN 0-8053-9049-9.
- Griffiths, D. J. (1999). Introduction to Electrodynamics (3rd ed.). Prentice Hall. pp. 301-303. ISBN 0-13-805326-X.
- Tipler, P. A.; Mosca, G. (2003). Physics for Scientists and Engineers (5th ed.). W.H. Freeman. p. 795. ISBN 978-0716708100.
- Jordan, E.; Balmain, K. G. (1968). Electromagnetic Waves and Radiating Systems (2nd ed.). Prentice-Hall. p. 100.
- Hayt, W. (1989). Engineering Electromagnetics (5th ed.). McGraw-Hill. p. 312. ISBN 0-07-027406-1.
- Schmitt, R. (2002). Electromagnetics Explained. p. 75.
- Whelan, P. M.; Hodgeson, M. J. (1978). Essential Principles of Physics (2nd ed.). John Murray. ISBN 0-7195-3382-1.
- Nave, C. R. "Faraday's Law". HyperPhysics. Georgia State University. Retrieved 2011-08-29.
- Maxwell, J. C. (1861). "On physical lines of force". Philosophical Magazine. 90: 11-23. doi:10.1080/1478643100365918.